BOAT TECH
by
BOB BERNSTEIN

Copyright © 1992-2014 Capt. Robert G. Bernstein

All Rights Reserved

Product Photos Courtesy of Product Manufacturers

ISBN-13: 978-1495941344
ISBN-10: 1495941345
ASIN: B00I8W2C9E

For Dad

FORWARD

These articles were written years ago while I was working as a commercial fisherman, passenger boat owner/operator, Field Editor for the country's premier commercial fishing publication, *National Fishermen*, and freelancer for several other national and international marine magazines. The work collected here is not meant to be all inclusive, and in no way does it cover the breadth of equipment and gear found on a given recreational or commercial boat. However, anyone looking to purchase new equipment or begin a new project, whether it's painting the hull or taking out an old engine and replacing with a new one, may find some useful, money-saving advice in these pages. Wherever possible, industry experts were consulted and interviewed, and most of the interviewees quoted here are still in the business.

Is this book worth $.99 as a Kindle download? Only the reader can answer that question. My advice is to look over the Table of Contents and seek out subjects that are of immediate concern to you. Are you planning to buy a watermaker, a pair of expensive binoculars? Are you considering a transatlantic passage? Do your navigation electronics make you feel safe and secure? How about upgrading your steering system, or putting in a 12 volt diesel furnace? Would you like to start a program of condition based maintenance for your boat or your fleet? Do you think you can install your own Single Side Band radio?

The main purpose for putting these articles together is to see if it will serve as an introduction to a much larger book on boat systems and boat gear. Hopefully, over time, readers will write in requests for information on the subjects that are of most interest to them. With the right kind of feedback, a second addition can be written that more throughly encompasses the needs of every recreational and commercial boater.

Thanks for ordering the book.

➢ Capt. Bob

RAYMARINE ROTARY DRIVE UNIT

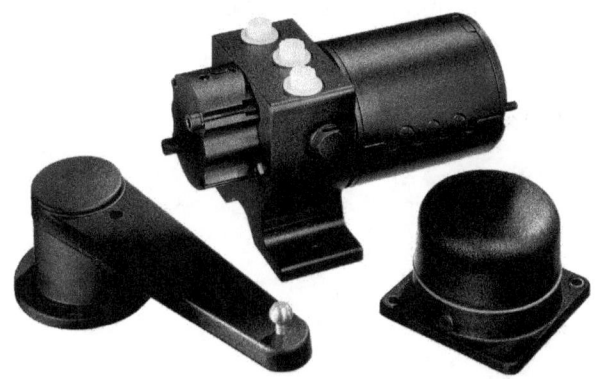

CHAPTER ONE
AUTOPILOTS

Keeping it Straight
For fishermen who don't have the money to gear up with everything from the start, the sequence of electronics purchases for their boats goes something like this: video sounder, GPS, radar, plotter...and autopilot. The autopilot is last. Why? Because when you're on a budget, the principle objective is to find the fish and bring them home...pronto! I suspect recreational boaters have a similar priority list, because for most mariners who aren't passagemakers, the autopilot is a luxury.

An autopilot won't help you catch fish. It won't help you navigate from Point A to Point B. It won't tell you where you are. It won't help you avoid underwater hazards or collisions at sea. It's even dumber than a rock when it comes to the Rules of the Road. And yet...everyone wants one.

Well, maybe not everyone.

Luther Blount, founder of Blount Marine in Warren, R.I., one of the country's most innovative shipbuilders, won't install an autopilot on any of his cruise boats. The reason? One time when he was at sea on a swordfish boat running on autopilot, he almost collided with a tanker. From that day forward, he swore, none of his boats would ever have an autopilot.

"I admit all the guys want it," said Blount to me many years ago. "Because it's like having another crewman. But I want the guy at the wheel. Even though a lot of people say I'm too conservative, I can go to sleep better knowing that someone's at the wheel."

Does this mean Blount would never run or own a boat with an autopilot? Not exactly. "If I was building a little yacht of my own, I'd put one on," he says.

Running Blind

"Any fisherman who says he hasn't fallen asleep in the wheelhouse with the ship on autopilot is lying," said a friend of mine who worked as chief engineer on an Alaskan factory trawler.

Running blind...this is the problem for fishermen and mariners in general. Does the boat you see on radar have a man at the helm, or is it steaming along on autopilot with no one on watch? Picture it: You're making a tow or setting gear at night and you see another boat headed in your direction, or you're at the helm of your sailboat and the radar alarm signals the proximity of another boat.

Boat Tech

You want to arrange passing signals, or just find out what the other guy's doing, so you try and hail him on VHF channel 16. No response. The boat gets closer and closer and still you get nothing back. Finally, you hit the horn. But on the other boat, the deck crew is busy changing nets on the net reel, the winches are wailing like banshees, the captain is in the galley, and the helmsman is fast asleep.

Jeff Ciampa, a Fishing Vessel Safety Specialist who used to be stationed at the Marine Safety Office in Portland, Maine knew the situation as well as anyone. "When we hear of a report of a grounding of an offshore fishing vessel," he said to me years ago, "and the vessel is westbound, returning to port, we try not to approach the incident with technical bias. But, generally, we find the problem to be a sleeping helmsman on a boat with throttle and autopilot engaged."

Ciampa said that fishing boats are more susceptible than other vessels to groundings and collisions due to the fact that they're sailing with minimal crews and deep fatigue issues. "For boats and crews fishing beyond 48 hours, fatigue is a real problem," he said.

Manufacturers are very well aware of fatigue and green crew problems, because boat owners and captains continually bring it to their attention.

One answer to this problem is the Helm Alert, a watch alarm with a buzzer that signals the helmsman to hit a button every so many minutes. If the helmsman neglects the button, the Helm Alert disables the autopilot and continues to sound its buzzer. It's not a foolproof solution to the manning problem, but it's better than nothing. Coupled with a radar alarm, it can give a nervous captain a couple of hours of peaceful sleep.

Meanwhile, as manufacturers are quick to point out, it's not the machines that cause these problems, it's the people who run them. In fact, a U.S. Coast Guard report bears this out. It found that as much as 80% of marine casualties are due to human error. Not really much of a surprise there.

Holding Her Steady

There are a lot of reasons why an autopilot will have a hard time steering a boat, some have to do with the boat itself, and some have to do with the autopilot, either an application problem or a malfunction.

For a lot of retailers and boat builders, the autopilot is one of the least favorite things to sell and spec out, but they admit it's a real accomplishment when they install one that works well. A lot of the time, they hear the same complaint. The autopilot is hunting.

Determining specifications for an autopilot is a complicated and involved process. You have to have a great deal of information, including details about rudder size, shape, taper, etc. And the drive devices and the sensors have to be compatible and tuned. For example, it's important that the components speak the same language. The heading sensor can't be sending data in 1 degree increments if the control unit is only capable of understanding 2 degree increments.

As far as the heading sensor is concerned: Provided it's compatible, most everyone agrees that for big vessels, the gyro is the way to go, but these are extremely expensive, $10,000 to $15,000 and up. After this, there's some difference of opinion. Some people are fond of the rate compensating or fluxgate Compass. Others recommend an electronic pickup and the largest and best magnetic compass you can get, usually with a minimum 6" diameter card.

So, the choices are electronic compasses, compasses with electronic pickups, fluxgate compasses, full blown gyros, and solid state gyros. The solid state gyros don't know where true north is, but they will do a good job steering the boat. And Fluxgates can have problems in steel boats. Basically, if money is no object, the best choice is the full blown gyro. If money is an issue, try the fluxgate first, and if it has problems, switch to a solid state gyro.

Meanwhile, a lot of fishermen blame their autopilot for a problem that's really not its fault. The truth is that even the best autopilot will have a hard time steering a boat that's inherently hard to steer. In fact, with more and more of today's vessels being built wider and fatter, with bigger and more powerful engines, autopilots have their jobs cut out for them. As one autopilot engineer explained it to me:

"You design a pilot for a 747, they all work the same. But boats are all different. Some boats steer very well, and some boats steer very badly. These days, designers are more concerned with how fast it goes, how much weight it carries, but not so much with how it steers. For example, take a Bristol Bay boat, they're almost boxes now."

This particular engineer works for a company that has thousands of autopilots in operation in Alaska. He told me there are a lot of factors that contribute to bad steering, including something called dead water, a consequence of a vessel's design. "This happens when a rudder isn't designed well enough to steer the boat," he told me.

It's no surprise then that a captain can learn how to steer a difficult boat better than any autopilot, because he's looking at waves, feeling the motion of the boat, anticipating the vessel's pitch and roll and compensating accordingly. An autopilot can't really do that.

Another problem is vessel trim. An autopilot may hold the boat's course perfectly when the vessel's full of fuel and ice, but when both are half gone and not replaced with fish, the boat might be bow heavy and plow or list to one side or the other. These changes make it harder for an autopilot to compensate. According to engineers, a lot of captains will try to correct the problem with the controls on the autopilot when they should be adjusting the vessel's trim instead.

Autopilots are more sophisticated than ever thanks to advances in sensing equipment, computer software, and the interfacing of plotters and GPS receivers. Today, you can set your autopilot directly from the plotter, run your boat along a prescribed route, and have the autopilot make course changes at each and every waypoint. Pretty amazing stuff when you think about it.

But this is just the beginning. In the years to come, autopilots coupled to thermal imaging radar, high speed computer plotters, sonars, and satellite transmitter/receivers will make it possible for you to automate every step of your trip, maybe even including docking an undocking. Would you want to some day be able to fish from the couch in your living room using a desktop computer and a couple of joysticks? Probably not, especially when you consider the safety issues discussed above.

On the other hand, with or without worrying about safety, you might not have 100% confidence in the computer that runs the whole show. Face it: There's always the occasional glitch that surfaces when you cross paths with a computer. And since autopilots (and other marine electronics) are relying more and more CPUs, the glitch event is something you just have to be ready for, and, in some cases, learn to live with.

This, of course, isn't to say that the new autopilots are unreliable. On the contrary, they're more accurate and dependable than ever before, particularly in terms of sensors. Look at the older photoelectric sensors. They had a much higher failure rate than the sensors we have today.

With autopilots, the basic components -- heading

sensor, control unit,, and drive device -- still do the same job they did twenty years ago, but their internal mechanisms and interfaces are much improved. Good thing, too, because the job they're being asked to do is getting harder and harder all the time.

FIN-NOR SANTIAGO

CHAPTER TWO
THE TUNA REEL

Catching the Big Ones
It has been said the giant bluefin has more power and stamina than any other fish in the ocean. As one arm weary angler put it, "It's like hooking up to a car going 50 mph."

In fact, the giant bluefin, largest and arguably strongest of all the bony fishes, can reach speeds of 25 mph, lengths of 14 feet, and weigh in at well over 1400 pounds. (The IGFA all tackle record, caught off Nova Scotia with rod and reel in 1979, was a fish weighing 1496 pounds.)

Scientists may one day calculate the actual maximum force one of these fish exerts. Until then, it's enough for fishermen to know that a heavyweight bluefin, with his head turned 180 degrees to the direction of the fight, and his tail pumping fast and furious, can part a 200 pound Dacron backing line with a report like that of a rifle shot. These fish are a formidable prey, and no match for the unprepared angler.

Add to this its value on the auction block -- $173,000 on January 5, 2001 for a 444 pound fish caught in the Pacific Ocean off Aomori Prefecture in northern Japan, one of the highest quality producing tuna grounds in the world -- and you have a potential catch that demands the utmost respect and attention to detail.

Strength in Numbers

The first recorded use of angling gear on a tuna was in 1871. Thomas Patillo, a school teacher from a small town in Nova Scotia, used a home made outfit and landed a 600 pound fish that had been cut free of a herring weir in Liverpool Harbour.

If he were alive today, Patillo would find reels a lot easier to come by, considerably more sophisticated than his crude contraption, and a lot more costly. However, one thing he'd find hadn't changed. It's still all about catching the fish.

There's one guy I know who does darn well catching tuna on a rod and reel, and he likes to use the old tried and true Penn 117. "In a smaller center console boat, 20' to 25', the 117 can do the job as well as the bigger reel in a larger boat," he says. "You have to work a lot harder with a 117, but it's about half the price of an International 80 [$400+ versus $800+] and it holds as much line."

From the point of view of dollars and cents, it's hard to argue with the man's logic. As he explains: You can buy twice the number of 117 reels for the money you'd pay for the 80s, or similar reels from another manufacturer, which means putting more bait in the water, which in turn means doubling your chances of catching fish.

What most fishermen do later, after they have a few fish under their belt, is they sell or trade their 117s and move up to the bigger, more expensive two-speed reels. Slowly, over time, they upgrade their arsenals.

A good comparison would be to the trap fisherman who starts out with used, rebuilt traps. Over time, as fishing improves and money flows more steadily, he phases out the old gear for new.

The Reel Market

In the previous example, I mentioned Penn reels in order to illustrate the type of buying one might do on a particular budget, but I'm the first to admit there are other alternatives to the tried and true. In fact, the Shimano Tiagra 130 is a very nice two-speed reel with what Shimanp calls it's hydrothermal drag system. They claim it's a smoother drag than a star drag because it eliminates setting variances. Personally, I find the lever drags more precise. However, as far as materials go, both Penn and Shimano stick to what works, and that's basically the same for both companies.

Of course, Penns are probably more widely known

than any other type of reel. The company has been in business longer than anyone else, one year longer than Fin-Nor, and few people will argue with the claim that Penns have caught more fish than any other type of reel. They hold the record for most number of IGFA registered fish caught, over 1100, and continue to lead the pack in sales. The general perception is one of proven performance and simplicity, particularly in terms of their drag system. In contrast, the Fin-Nor reel -- with a history dating back to Ernest Hemingway's penning of The Old Man and the Sea—it was a Fin-Nor that caught the 1560 pound black marlin for the film version of the author's famous ode to fishermen—has a more complex drag design. Taking one apart is like taking apart a Swiss watch.

Fin-Nor had stopped making the big game 80 lb. and 130 lb. class trolling reels. For awhile, the last catalog that included them was the 1999 catalogue. However, the company, now located in Wilmington, NC, always maintained a large inventory of parts, and for years the talk was they would be bringing both classes of reel back. Sure enough they did, both the 80 and the 130, with prices around $900 and $1300 respectively.

Parts for old Fin-Nor reels can be a bit tricky to get. Pawl springs are fairly easy to come by, but finding early model spool shafts can be tougher. The drag plate and spool shaft in the early models were machined out of one piece of aluminum. Newer models include a two piece spool shaft and drag plate.

At Tight Lines Tackle, in Walpole, Maine, where they

sell Penn, Shimano, and Everol, owner Dave Mason adds his voice to the chorus of industry people who rate Penn as the leader in terms of popularity among commercial fisherman. But, says Mason, Everol has been making an increasingly strong showing. He also says Fin-Nor is a top-rated reel with a loyal following. When asked if the Fin-Nor is more of the sportsman's choice, he answers: "When it comes to tuna, everybody's commercial."

Mason also agrees with my old fisherman friend about the Shimano 130. "I had a chance to use one last year and caught a 900 pound fish. I was very impressed with the reel."

Big game Fin-Nor trolling reels, both two-speed and three-speed models, new and used, are still available from tackle shops and on-line used equipment dealers. An antique auctioneer told me he gets calls all the time from fishermen asking what their reels are worth.

Meanwhile, the nine companies competing in the world of two-speed 80 lb. class and up category reels are: Accurate (USA), Alutecnos (Italy), Daiwa (Japan), Duel (Italy), Everol (Italy), Fin-Nor (USA), Okuma (Japan), Penn (USA), and Shimano (Japan). For a 130 lb. class two-speed reel, prices range from about $850.00 for the Everol, to about $2500.00 for the Accurate. Penns, Alutecnos and Shimanos run in the neighborhood of $1200.00. The top of the line Duel costs $1800.00. Neither Daiwa or Okuma list a 130 lb. class reel. (Prices are based on old data and are probably 30% more today.)

All things considered, it's the drag that sets different makes of reels apart. "You need a smooth drag," says one fisherman. "Tuna are very strong. If you don't have a smooth drag, you'll end up with a very jerky type of play. Too much of that and you'll part line."

Unlike other game fish (marlin, sailfish, mako), a giant bluefin, at least the Atlantic variety, doesn't tire itself out with wasted motion. It doesn't try to shake a hook loose in the air or get into fanciful displays of acrobatics. For the most part, the fish turns and runs using powerful strokes of its tail. This type of play will heat things up in a hurry.

"Drag washers have facings on them like a car brake," says Mason. "When they get heated up, parts of the washer can liquefy."

All big game reels now use a lever drag system consisting of a clutch made up of multiple discs. This system is designed to let the line out in a smooth, controlled manner when the fish pulls.

Accurate claims to have the only twin drag system, in which both sides of the reel are fixed with drag plates and washers. Shimano boasts a double drag plate system. And Everol claims to have the only variable lever drag system.

On the Everol reel, a graduated side plate allows a fisherman to preset the drag without using a scale.

In this writer's view, Everol is daring fishermen to do more in terms of playing the fish with the drag, while other companies prefer to stick to the more conventional methods of angling, vis a vis presetting the strike drag with a scale and keeping a more hands off approach to the drag during a fight.

"Usually, anglers will use a Cardoza Creation scale to set the reels at 32 lbs. to 35 lbs at strike. And they generally don't move the drag again until they're ready to harpoon, then then bump the drag up over strike to prevent the fish from taking ten or twenty feet of line out.

It's a matter of physics. As line leaves the reel and the diameter of the spool decreases, the drag force increases. As spool diameter increases, drag force decreases. And if there's any one common mistake novice anglers make, it's using too much drag."

Clearly, given the stakes, setting and maintaining proper drag is up there in importance with being in the right place at the right time and presenting a sufficiently tempting bait. Perhaps the next thing is being there with a reel you trust and appreciate.

"It's funny," says one sport fishing dealer I know, "I call these doctors and high power lawyers, and their nurses or secretaries will tell me he's in with a patient or a client. They'll ask me what the call's about and I'll say I'm calling about their reel. The next thing I know they're saying, ' Oh, hold on, he'll be right with you."

MY BOAT PAINTED AND READY TO LAUNCH

CHAPTER THREE
BOTTOM PAINT

Bottom Paint

If you don't take the time preparing the bottom of a new boat, you're bound to end up with big patches of peeling paint. Mold release agents and wax are the primary culprits for new fiberglass boats, and it's not easy to get them off. Paint manufacturers recommend the following: Wash the boat, then wipe it down with a cheesecloth dampened in solvent. Then, after the entire bottom has been cleaned with solvent, spray the hull with water to see if you missed any spots; water will bead-up on wax. Go over the spots you missed, then wash it again, repeating the process until no more water beads-up on the hull. Finally, after the hull has been completely cleaned of wax, sand it with very fine sand paper or a Scotch Brite Pad, because a little scuffing gives the paint a surface it can adhere to.

Using the two handed method when wiping down the hull works well, one hand on the damp cloth and the other on a dry cloth. Apply the solvent, then remove it right away with the dry cloth. And replace the rags (cheesecloth, pads or whatever) as frequently as possible. You can't have any wax left on the boat at all. If you do, you'll just end up playing catch up for the next few years.

Wood, aluminum, and steel boats are a little different. But in each case, taking the time to prepare the hull is just as important.

For bare wood, you should have a clean dry surface and a primer coat of bottom paint mixed with thinner. Lay the primer coat on, then fill the seams with seam compound, then put on your bottom paint at normal strength. Thin the primer 5% to 15% and let it dry overnight. Oily woods like mahogany and long leaf yellow pine should be allowed to season as long as possible. Green wood and wood with knots (or rather the knots themselves) will reject paint no matter what you do.

Aluminum and steel are a bit more complicated.

Painting Metal Hulls with Anti-foulant

There's a huge balancing act that takes place when you paint the bottom of a metal boat or its shafts, through-hulls, props, rudders, etc. The reason for this is that bottom paint has a lot of copper in it, and copper is higher on the galvanic scale than steel and aluminum. So, if you're not careful, any exposed steel or aluminum, and even the steel skeg or a shoe on a wooden boat, will slowly give itself up to its coating. In other words, it will start corroding.

Even though an aluminum or steel hull needs a barrier coat between the anti-fouling paint and the metal, the preparation of the hull is pretty much the same as it is for fiberglass. Instead of wax, you're removing grease and other thin film contaminants. So, with a new steel or aluminum hull: wash, treat chemically, wash again, scuff (either by sandblasting or with medium grit emery cloth), and repeat as needed to remove grease and film. Then clean the hull with a primer wash (usually a diluted solvent), wait a certain period for it to dry, but not too long that contaminants begin to adhere to the hull again, then apply the barrier coat. Don't skimp on the barrier. Use two to four coats to guarantee adequate coverage. Remember, this is all that stands between your hull and the corrosive effects of the anti-fouling paint.

Meanwhile, steel and aluminum hulls are not the only metals that have to be protected. A case in point: Zebra mussels. These bivalves have become a serious problem around the Florida and Gulf coasts. As a result, owners have been painting running gear and through-hulls for years. Without anti-fouling protection, a southern boat's propeller, trim tabs, shafts, rudders, etc. will attract mussels. All you need is one mussel to attache to an exposed surface. After that, one will grow one on top of the other and expand into a colony like a string of garlic.

But even in the northern. colder climes, more and more people are painting their shafts, props, and struts. In some cases, they do this to avoid galvanic corrosion. "We paint everything," says George Emory of the Tenants Harbor Boat Shop in Tenants Harbor, Maine.

"For example, in a wooden boat, if you put too much zinc on, you dissolve the wood. But we find that if you paint a manganese bronze casting with a synthetic ' red lead', you avoid a lot of galvanic interaction and the need to use too many zincs. So we prime with ' red lead', then paint with what the owner wants for antifoulant."

As far as the schedule for priming a steel boat goes, Jack Brady, a Paint Foreman for the Lake Union Dry Dock Company in Seattle, Washington, says that he and his crew put the first coat of primer on the first day, the second coat on the second day, then—after the second coat of primer's tacky, passing what Brady calls the thumb print test—they lay on the anti-fouling paint. "We use two coats of primer," says Brady, "and it usually lasts five to seven years, except where there's a lot of chafing, as with the crab boats. They come in here, and they're always damaged from hauling or from ice. They take a lot of abuse."

Taking off Old Paint

Taking off old anti-fouling paint is one of the worst, if not the worst, job around. Whereas the fumes from new anti-fouling paint can be harmful if inhaled, the dust from old paint poses a much more serious health risk in terms of lung disease. And yet time and time again, yard owners and foreman see fishermen sanding their hulls and not wearing adequate protection. It's not a smart thing to do, and these days, depending on where you are and who's watching, it can be a violation of the law and a costly error in judgement. If you're going to do it, wear protective gear, and use a vacuum sander.

For fiberglass and wood, the preferred method of taking off old paint is to pressure wash, then sand. But sanding is an incredibly messy and potentially risky process, so it pays to heed the advice of the experts and follow the OSHA warnings on the labels. By all means, use a fully self-contained sanding system that includes the vacuum sander, the hose, and the recovery container— and, at the very least, wear protective masks and clothing.

Does accumulated paint have to come off every year? It depends on the paint...and the boat's service. Brady reports that most of the fish boats that come into the Lake Union Dry Dock Company after being out of the water for two years or less look pretty good; the yard can clean them with a 3,000 psi pressure washer. If a boat comes in that has worn through some of its paint and primer, e.g. a crab boat or other vessel that fishes through winter ice, the crew will take the area down to bare metal, either by sandblasting or ultra-high pressure washing.

Ultra-high pressure washing (30,000 psi and up) is am effective method of surface cleaning. It has the advantage of being significantly more environmentally friendly than older methods of abrasive or slurry (a mix of abrasives and water) blasting, and it doesn't create airborne pollutants. It doesn't need to be recovered in the same way, either; whereas the sand or other type of blasting material has to be contained and collected, the majority of the water from an ultra-high pressure water-jet evaporates on contact. Only the surface material (paint, rust, and other contaminants) are left to be collected.

At 40,000 psi, the nozzle of an ultra-high pressure washer can be controlled by one man. At higher pressures and flow rates, robot controllers are needed. Flow International Corporation, in Kent Washington, manufactures several different types of ultra-high pressure water-jets, including the Husky(TM) Water-jet System that operates with water flow rates of less than 2.5 gpm. Compare this to the 20 to 30 gpm used by a typical low pressure washer at 3,000 psi.

For these reasons—dust free, low water flow rates, no need for containment or the recovery and transport of abrasives—the industry seems to be swinging in the water-jet direction. At Todd Pacific Shipyards in Seattle, Washington, the majority of owners used to choose abrasive blasting over any other type of surface preparation method. However, since the yard introduced the Husky Water-jet System in 1994, word is spreading fast. According to former Pacific Shipyard's Manager, Allen D. Rainsberger (now the Manager of the Environmental Coalition of South Seattle) boat owners choose four to one in favor of ultra-high pressure washing.

This isn't to say that the ultra-high pressure preparation method is the perfect solution. There are downsides. For one, there's the fact that many primer and barrier coats can't stand up to the force of the spray. Also, the procedure tends to discolor the surface of metal, leaving behind what many people consider the initial stages of oxidation. And, as Brady points out: If you're not careful, the jet it'll knock off a transducer. "It'll take your foot off in a heartbeat," he adds.

Still, except for the pressures, which are part and parcel of the jet's design, these problems are being addressed by paint manufacturers and engineers and chemists at Flow International. Among other things, they're experimenting with different primers and barrier coats and looking at a non-toxic rust inhibitor to add to the water.

Meanwhile, whether you clean by virtue of abrasive blasting or ultra-high pressure washing, sand with a vacuum power sander or a piece of medium grit emory cloth, wipe with cheesecloth or a rag, the objective is the same: Prepare the bottom in the best way possible, and the perfect application of bottom paint will follow.

In addition, check to make sure you've researched the latest advancements in anti-fouling paint. There are many new products on the market today, and many of them require stripping the hull to bare glass or metal before application.

ICOM M802 SSB

CHAPTER FOUR
SINGLE SIDE BAND

Counterpoise is King
Of the four most popular marine communications systems in use today—VHF, SATCOM, CELLULAR, and SINGLE SIDE BAND, the latter, the SSB is the most challenging to install. Does this mean you should hire a professional technician to put it in? Not necessarily.

According to the experts, recreational and commercial boaters don't always have to hire a technician when it comes to installing their single side band systems. However, they should have a professional on hand to provide consultation and advice. "Seafarers are an independent-minded bunch," says one engineer and SSB installation specialist I spoke to, "and they know their equipment and machinery. They have to. If something breaks down when they're at sea, they're the ones who have to fix it. Some of them, like commercial fishermen, are also cost-conscious. So I understand why many of them have a ' do it yourself' attitude. In fact, we talk people through these isntalls all the time."

That said, it's a given that some of these installations can be quite tricky. The biggest mistake people make is in underestimating the need for a suitable counterpoise.

Counterpoise, by definition, is an equal and opposing power or force, or something that creates balance or a state of equilibrium. Picture a springboard. The counterpoise would be the anchor or counterweight in the springboard's base. Without the counterweight, a person jumping on the board would probably end up flat on his face.

Same thing happens with a SSB. The counterpoise acts as the base of an electromagnetic springboard, and it's from this that the transmission gets "boosted" off the boat and into the atmosphere. In other words, without it, you, or more appropriately your communication, won't get very far.

Another way to look at it is as follows: A SSB antenna is not just the piece of fiberglass encased metal or length of wire mounted to the top of a boat. It's everything from the bottom of the system up. So, for the same reason you wouldn't install half a VHF antenna, you wouldn't install a SSB without a suitable counterpoise.

Counterpoise

In terms of SSB communications, counterpoise comes from a good ground. On land, this would be a wire soldered to a pipe and driven into the earth. On the water, it's a bit more complicated.

Basically, there are four ways to establish counterpoise on a boat:

(1) Using the ocean as a ground.

(2) Using the vessel's machinery and equipment as a ground.

(3) Combining both one and two.

(4) Using a ground plane, which, from a purely technical standpoint, is a sheet of conducting material that is at zero potential and provides a low impedance earth connection across its entire length.

For example: Because it has no shortage of good conducting material, a steel boat provides an excellent connection to the water. This means that the coupler—an intermediate device that ties together the antenna, the transmitter and the connection to the ground—can be installed anywhere on the roof of the wheelhouse. Mounted this way, the entire hull of the boat at the waterline serves as a ground plane.

Non-metal boats, however, are another story, because when you don't have a enough metal in contact with the water, you have to find a different solution. Sometimes that solution can be found by bonding the vessel's machinery and mechanical components together, engines, generators, exhaust stacks, fuel tanks, through hulls, plumbing, etc. Other times it can be found by bonding the machinery together and tying them into a connection to the water, e.g. the shafts. Another choice is to bed a swath of copper mesh into the roof of the wheelhouse. How much of this should you use? A minimum of 100 square feet, a 10' x 10', is what most professionals square will recommend.

Bill Tener (pronounced "Teener"), Owner of South Central Radar in Homer, AK, can remember a time when fishermen could be seen walking down the wharf with bales of brass screen wire under their arms. But this, he says, was before the microprocessor controlled coupler. These days, Tener says, he has very good luck establishing a ground in non-metal boats by getting the antenna coupler as close to the water as possible, which means mounting it in the engine room. As he explains: "The new matching antenna control devices [another name for the coupler] seem to handle the limited ground systems better."

Still, Tener admits that he sometimes runs into other problems with this type of installation. "Occasionally, when the coupler is mounted too close to a [boat's] charging circuit, we get 'noise.'

Noise is a problem because radios can't differentiate between the signal they're supposed to be receiving and the interference. What happens is that the automatic squelch cuts-in when it thinks it's getting a good signal. But what it's really getting is noise from things like the alternator, pump motors, voltage regulators, ignition systems, etc. However, these problems are easily solved.

The various sources are tracked down, and the culprits are either shielded, bonded, or filtered. "It can be tough chasing down some of these problems," says Tener. "But in order to pass a FCC or Coast Guard inspection, you have to get these things quieted down. Some of the inspectors can be very helpful with this, but either way," and here he echoes the prevailing opinion when problems on a boat get tough to figure out, "you really should have someone around who know's what he's doing."

Other Considerations

"Once every two years," Tener says, "you should check the resistance of the antenna. You should also keep the insulators clean. Stack smoke gets on the insulators, or salt; they get sticky, and then they lose their insulative properties. We have people who paint the insulators." Obviously, this is not a good idea."

Another thing you want to check is the mounting of the antenna itself. These are fairly large antennas to start with, and this makes them susceptible to breakage in heavy seas. Keep them secure and as protected as is humanly possible. Tener also recommends carrying an emergency antenna system or at least being prepared to jury rig one off of the vessel's stays. The reason for this is that if you're dismasted during an emergency all you need to do is have the proper length and type of wire pre-made and ready to go. You could spread it out on deck if you have to.

Choosing the right radio and antenna is critical,

because there are many different types of radios and antennas. What you choose will depend on your communications needs and your boat. For example, a large sailboat can use a high voltage wire with two insulators at either end as a backstay. But these are more expensive and not always practical. The norm is just what you'd expect. If you're trying to talk 400 to 500 miles, a practical solution would be a 150 watt radio, and a Morad antenna. This combination can work very well in this range."

If you're hearing about counterpoise, ground planes, and antenna couplers for the first time, chances are you're going to need some help with an installation. This is an important point to remember because service representatives—and who can blame them—don't like to support equipment they don't sell. In fact, for this very reason, Tener and others, seriously question the advisability of buying directly from a catalogue. Something to think about as you plan your purchase.

SCHEMATIC OF EBI TC20

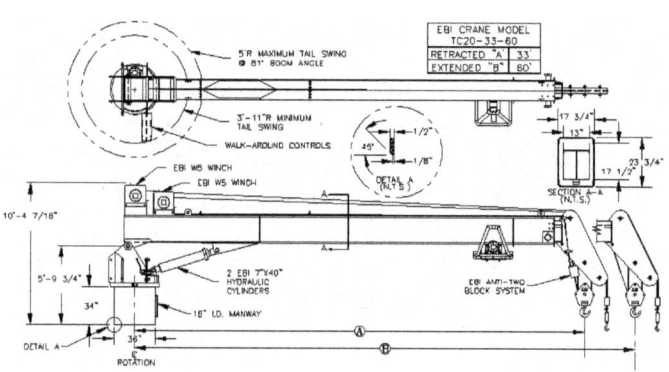

CHAPTER FIVE
CRANES

How Much Can You Lift?

A hundred and fifty years ago there wasn't much choice when it came to hoisting something aboard. You had a yard arm and a system of ropes and pulleys and not much else. Fishermen would saddle themselves to a length of cordage and heave-ho in unison, maybe to the tune of a 19th century Scandinavian hauling song like, "Halarvisa":

> [chorus]
> Karre, varre, vitt bom bom!
> Nicke, dicke, dickum, plutt!

Fortunately, the typical hand-over-hand team work of the past has been replaced by solo crane operators manipulating joy sticks in heated or air-conditioned control cabs. And hauling "sing-outs" like the one above have given way to headphones and digital audio CDs.

Well, maybe not all hoisting or lifting operations are that luxurious, but there's no debating the fact that today's deck cranes have made life on fish boats a whole lot easier and safer.

Choosing a Crane

Basically, here are your different types:

(1) The Fixed Boom Crane is the simplest. It's a three function crane that can boom (luff), swing (slue), and winch (hoist).

(2) The Telescopic or Extension Boom Crane is a four function crane that can boom, swing, winch, and extend.

(3) The Knuckle Boom Crane, or Articulated Crane, is another four function crane that can boom, swing, winch, and, as the name implies,"knuckle."

(4) The Knuckle-Extension Crane is a five function crane that can boom, knuckle, winch, swing, and extend.

A further bit of clarification comes from William Morgan, Owner of The Morgan Company in Fortuna, California, fabricators and distributors of marine cranes up to and including offshore cranes with capacities of 1,000,000 foot-pounds and telescopic and knuckle boom cranes up to 300,000 foot-pounds. According to Morgan, the true articulated crane folds into a figure four. "Many knuckle or articulated cranes may not fold completely into a figure four and not be true articulated cranes," he says.

Whether you choose, fixed, telescoping, or articulated

depends a lot on the application and the fishery, but also on personal preference and how much versatility you need. For example, while you're likely to find knuckle cranes on the majority of Alaskan crab boats and extension cranes on most factory trawlers, Washburn insists, "There's no hard and fast rule." Trawlers will use knuckle cranes, and some crabbers will opt for extension cranes.

"Crabbers prefer the knuckle or knuckle-extension crane," says Washburn, "for the simple reason that it gives them more control of the pot. "The knuckle crane," he explains, "allows the operator to keep the tip of the crane closer to the load, which makes it safer on a crab boat. On a crab boat you [also] have a pot guard or a crab cage at the tip of the boom. It's a kind of a tripod arrangement with a rubber tire that allows you to compress the tire with the pot right at the sheave head."

Knuckle or articulated cranes have a several other advantages. True articulated cranes can fold up and stow in less deck space, and, in general, knuckle cranes are faster to operate. As William Morgan explains: "A telescoping crane is usually operating in one plane, one function at a time. They're very slow because the operator winches, then booms, then swings. Articulated cranes are much faster. They can move in three planes at once, like an extension of your arm."

Morgan also points out that the articulated crane is the highest capacity crane for its size, weight, and cost. "A telescoping crane runs at lower pressure, so it has larger cylinders," he says. "The knuckle crane has higher pressures and smaller cylinders, and it's made of higher strength steel. Overall it's a lighter crane and it lifts more."

The fact that Morgan says the articulated crane is less expensive will probably come as a surprise to many people shopping for their first crane. "You take the same capacity crane—telescoping versus articulated—and your talking up to twenty five percent less money for the articulating crane," says Morgan.

The reason for this, as he explains, is that manufactures build more articulated cranes than telescoping cranes. In addition, the majority of telescoping cranes use a ball bearing for rotation. "Our articulated cranes," says Morgan, "use a rack and pinion system for rotation. The rack and pinion, which actually has higher torque, costs less."

Rack and pinion systems are also used to extend the booms in some telescoping cranes. "The only cylinders we have [in our cranes] are the lift cylinders," says Mike Atkins, former Sales Manager for The Techcrane Global Corporation in Covington, LA, manufacturers representatives for EBI fixed boom and telescoping cranes. "We use lift cylinders," he explains, "but instead of an internal extension cylinder—the part that extends the boom in and out on most all other cranes—ours have a rack and pinion system. This eliminates any reason to get inside the boom for maintenance or repair, for example, if you had a broken seal. With our system you never have to get internal to repair a leak."

Meanwhile, regardless of each crane companies proprietary engineering (and there are many more crane manufacturers and distributors than those contacted for this article), there's no question that of the four generic types of cranes available, the knuckle or articulated is the most complicated. It's structurally more complex and also more complex to operate. "You need an operator who can think [faster] because the crane is operating in three planes," says Morgan.

What this means is that not every owner or captain wants the additional complexity of an articulated crane. For example, a factory trawler, where deck space may not be at a premium, may only need a crane to launch and recover a tender in and out of a fixed cradle. Or maybe the crane is needed to move nets on and off the boat two or three times a year. In cases like these, the relative simplicity and ease of operation of a fixed boom or telescoping crane might be more appealing.

Running the Numbers

Once you've picked a type of crane, the next step is to spec it out. In this regard, the most important variable is what Washburn refers to as the "Defining Moment."

"The first thing you have to do is figure out how much you want to lift and at what radius," says Washburn. "Those are the determining factors. Lift and radius."

The particular 'moment' Washburn is talking about is a quantity defined as the cross product of linear force and the distance from the point of rotation at which that force is applied, otherwise known as torque. Torque, or the moment of force, is usually defined in foot-pounds or tons-meters.

For example, say you want to lift 3000 pounds at 50' and 20,000 pounds at 20'. You multiply 50' x 3000 lbs., and 20' x 20,000 lbs., and you come up with moments of 150,000 foot-pounds and 400,000 foot-pounds respectively. Clearly, the defining moment is the one for which the product is 400,000 foot-pounds. That's the one the crane has to be designed and built for.

"The defining moment determines the bearing and the pedestal," says Washburn."[Here at Hydra-Pro] we deal in Single Row Ball Cranes. This [design consists] of a single row of balls that ride in a race mounted in between the pedestal and the turret."

The turret and pedestal are where most of the stress and fatigue will manifest. In fact. according to a recent Crane Safety Workshop conducted by the Offshore Technology Research Center and the Minerals Management Service of the U.S. Department of the Interior, the majority of marine crane failure incidents are pedestal related. This is why the design and construction of the pedestal and turret of a marine crane are so important.

On land, the lift is static, but at sea, where the vessel is moving side to side and fore and aft, the lift is dynamic. For example, an empty crab pot weighs about 700 pounds. Loaded with product it might weigh as much as 1300 pounds. With the vessel rolling and pitching, that weight can increase thirty to thirty five percent, adding another 450 pounds to the pot. However, as Washburn points out, the dynamic doesn't only affect the load: "The dynamic acts on the entire crane. You've got the weight of everything to consider, crane structure as well."

According to Washburn, normal accelerations will be at a dynamic of less than 1.33, i.e. 33% over static load. (1.0 is considered zero dynamic.) Hydra Pro cranes, and cranes from other established marine crane companies, are designed with a minimum of 1.33 dynamic. "Cranes built with 1.33 to 2.0 dynamics are relatively common," says Washburn. Although, he adds, Hydra Pro, as well as other companies, often supply cranes that meet or exceed specifications set forth by national and foreign regulatory agencies.

Morgan cranes are constructed by PM of Italy to meet DIN Standard 15081H1B2. "It's a German engineering standard," says Morgan. "It specifies a dynamic of over 3.0, and [includes standards for] how long a crane should last and how it should be built."

API (American Petroleum Institute), U.S.C.G., U.S. Navy, ABS (American Bureau of Shipping), Lloyds, Rina, Det Norske Veritas, and OSHA, are some of the other entities that establish (or share) standards for marine cranes. You may note that your crane meets API-2C specifications, which states, among other things, that the static load is 1.5 times the dynamic load.

Of course, you can get whatever you want. If you want to lift a dynamic load of 6000 pounds, and you want a 3.5 dynamic, you're going to get a crane that can lift a static load of 21,000 pounds. "We had a customer who requested a 3.5 dynamic," says Washburn. "He had an offshore application, and he had damaged three cranes in previous situations. These [the old cranes] had been cranes designed and built for shore applications."

Knowing how and when to conduct a given lift is where the load chart comes into play. In fact, in order to be compliant with regulatory agency standards, every crane on every vessel should have load/radius charts for static and dynamic lifting posted in plain sight of the operator.

The EBI TC10-24-40 telescoping boom is 24' in retracted mode and 40' in extended mode. A quick glance at the load chart shows it has a static lift capacity at 20' of 9750 pounds and a dynamic load capacity at 20' of 7200 pounds. The static load is 1.5 times the dynamic load.

Does this mean an operator has carte blanche to lift 7200 pounds in extended mode in any sea state? The answer to that is a resounding, no. Because you can't just make a blanket statement that this particular crane can lift 7200 pounds in any sea state. There are a lot of factors involved, and a lot of it has to do with the stability of the vessel. In addition to knowing the crane's operational parameters, you have to know what the vessel is capable of, and for that you need to check with a naval architect.

USCG PHOTO OF BOAT FIRE

CHAPTER SIX
ELECTRICAL

Wiring and Troubleshooting
What happened aboard a 50' Maine-based fishing boat not too long ago is a good example of why electrical problems shouldn't go untended. An electrical leak caused the vessel's two 8-D batteries to fall below 50% capacity. There was enough cranking power to get the engine started, but not much more. Consequently, the alternator had its work cut out for it just keeping the status quo, e.g. running the radar, depth sounders, Loran, lights, bilge pumps, etc.

On the return leg of a long steam to and from one of the local banks, the needle on the voltmeter started dropping. When the voltage got to below 12.6 volts the radar went out. Then the bilge pumps quit, followed by the running lights, cabin lights, and navigation electronics. Fortunately, shutting off everything but the running lights and radio enabled the crew to get home without further risk or incident. However, the next day, the batteries were completely flat and unusable.

All things considered, a short circuit or electrical leak to ground like what happened in the above account is relatively harmless. The cause of the difficulty was isolated and corrected, and the batteries replaced. (In most cases, particularly with starting or cranking batteries, a complete discharge spells a premature death and the best course of action is to replace them.) However, a situation like this can and often does cascade into something much worse, because small electrical shorts and leaks are a critical link in what safety professionals refer to as a "Risk Chain." The more links in the Risk Chain, the greater the chance for catastrophic failures. It might happen something like this:

Fog sets in, wind increases, radio traffic indicates lots of boats and ships in the sea lanes. Without radar, Loran or GPS, due to the loss of electrical power, the captain decides to shoot a course over shoal bottom, in his mind, taking the boat out of the traffic lanes and out of harms way. But misjudging the set and drift of the current, he hits a ledge and starts taking on water (or maybe stray current from the leak has corroded the fastenings around a plank and it lets go at an inopportune time). Now the boat has no electric bilge pumps and no lights, and the crew has no ability to radio a 25 Watt VHF distress call. So to contain the flooding, they try to disconnect the engine intake to use the engine as an emergency bilge pump, but the through hull breaks from stray current corrosion, causing an even worse flooding situation. Within minutes, they're facing an abandon ship scenario.

Visual Inspections

First and foremost, a vessel should be wired as cleanly and simply as possible, and only the right type of wire or cable should be used. According to Part 129 of CFR 46 for Shipping set forth by the U.S.C.G. and the Dept. of Transportation, cable must be listed by Underwriters Laboratories Inc. as "UL Boat Cable" or "UL Marine Shipboard Cable," or cable and wire must meet all construction and identification requirements of IEEE Std 45.

Basically, what these regulations refer to is C.G. Approved ' tinned' wire. Tinned wire is designed to be more resistant to corrosion than regular copper wire, and this is particularly important at the ends of the wire where connections are made, because this is where the conductor is most exposed. Tinned wire's greater flexibility makes it a lot easier to use, too.

Any boat that has substandard wire or cable and a poorly installed and maintained electrical system warrants a red flag. Suspicions should arise when ad hoc solutions to problems are visible. For example, what some people do when they notice a loss of DC power is grab a spare battery from the barn and add it to the bank they already have. This is the worst thing they can do. The oldest, weakest battery will drain the newer ones in no time at all. With batteries, everything eventually reduces to the lowest common denominator.

Other red flag areas are connections. Two bare wires twisted together, especially near the bilge, are an invitation to trouble. Nobody can say for sure how long electrical cable or wire will last in a boat, but Coast Guard Marine Safety Inspectors say it can have a life expectancy anywhere from five to twenty years. It all depends on where it's located, i.e. near heat and/or moisture like in the engine room next to an exhaust outlet, a tough environment for wire, or under the helm in a dry spot, a good place for wire. UV light and heat break down insulation, while vibration fatigues copper conductors and moisture corrodes them. And two bare wires twisted together will do almost as much to reduce the life expectancy of wire as storing it in a bucket of seawater.

For one thing, multi-strand wire is actually quite porous. Even the best will wick water like a sponge, which is why airtight connections are so important. Today, most builders use quality connectors and heat shrink tubing to prevent water corrosion and damage. They're also careful to choose the right size terminal or connector, and they use liquid electrical tape or dielectric coatings at exposed connections. Some builders will even take the wire leading to a terminal block, fuse block, or breaker panel, and bend it to form a "drip loop." The loop, which is bent into the wire before the connector, keeps the water from seeping onto the block or panel.

Of course, the best way to extend the life of your wire is to deal with corrosion and water damage as soon as it rears its ugly head. Green connectors are a sure sign of trouble and should be replaced as soon as possible. Just cut the wire back to clean conductor and put on a new connector. If, however, when you cut the wire back, all you find is brittle and rotted conductor, your only recourse is to run a whole new length of cable.

Trouble Shooting

When voltage drops and equipment starts failing, it's time to look for the source of the problem, and every mariner knows this can be like searching for the Holy Grail. But it has to be done, because if you have an unwanted open circuit, short circuit, or ground fault (leak to ground), the consequences can be expensive and potentially serious.

Here's where the multimeter becomes an indispensable tool. These hand-held gauges cost anywhere from $10 to over $400, with the latter being shockproof, waterproof, digital, and very accurate. But unless you're a professional electrician, the lower cost multimeter works just fine. And no boat should be without one or two. Most people can get all they need from a $10 multimeter, although some don't register amps, and having that capability to measure current is nice if you're concerned about a draw on the battery."

Measuring amperage is a good way to find out of you have a battery drain because an open circuit shouldn't have any current running through it. However, when using a multimeter, bear in mind that the majority have a range of only 250 milliamps. If you're not careful, you'll blow the meter. So, do not use this method to test for current leaks if you suspect the leak is going to be greater than the capacity of your multimeter. And do not do this with a freshly charged battery.

First thing to do is turn off all equipment using the on/off switch on the equipment. Then take the positive lead off the battery and put it back on. If it sparks, you've got a leak. If you have a visible spark in daylight, you can have as much as 5 amp leak, and maybe even more. A big spark means you have a major leak that you can probably find with a visual inspection.

[CAUTION: As everyone knows, batteries give off explosive gasses, particularly when charging and discharging. So take precautions. Engine rooms should be well ventilated and batteries should be capped. Also, make sure no piece of equipment will load the system when you're testing it. Bilge pumps and refrigerators will draw 5 to 10 amps each, more than enough to cause a good-sized spark.]

But let's say it's a small spark. Now it's time to locate the source. If you can't find it from a visual inspection, e.g. there's a corroded pump connection sitting in the bilge, then you'll have no choice but to run a test of the equipment through the main distribution panel, bearing in mind that some systems might don't go through the main panel, systems such as electric bilge pumps, engine ignition switches, engine gauges, and the alternator charging circuit; Note: it's not entirely impossible for there to be a leak going from the alternator into the engine block, a potentially costly place to have a short or ground fault.

Anyway, the systems that bypass the panel will have to

be tested individually. If they pass the test, and you still have the leak, disconnect one of the battery leads from its post and take the multimeter probes and connect them between the post and disconnected battery cable. Now go back to the panel and disconnect leads one at a time, or pull fuses, or turn off circuit breakers. If the current reading on the multimeter drops, then you've found the source of one of the leaks. If the problem still hasn't been found after this, the leak may be in the distribution panel itself, perhaps in one of the breakers. In which case, you'll have to test each of the breakers or the panel with the ohmmeter.

Make sure that while you're conducting the above test you don't have something automatically cycle on, such as a bilge pump triggered by the float switch, or the DC refrigerator turning on. These are high current devices that will surely blow most low cost multimeters.

Lastly, it should be noted that sometimes you can spend a lot of time searching for something that isn't a big deal, like the wasted effort spent finding out that a small amount of electricity is being used to move the hands on your ship's clock.

ROCK AND ROLL, BABY!

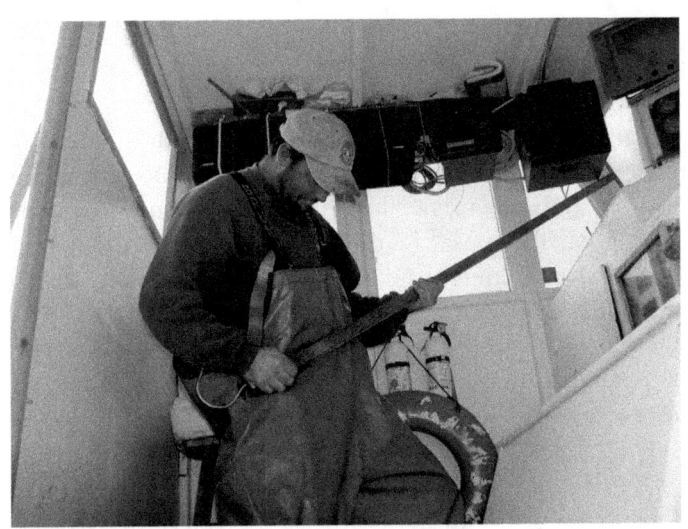

CHAPTER SEVEN
ENTERTAINMENT AT SEA

Stereos

It's not every fisherman who wants to listen to music while tending lines or hauling traps. It can be distracting to some and downright unsafe for others. And no matter how good the music is, there will always be those who prefer the droning sounds of the engine and sea to anything broadcast over the radio waves or recorded on CD or flash drive. (Personally, I'd rather hear the cries of herring gulls mating than be subjected to some of today's music, but that's neither here nor there.)

But let's say you're not one of those mariners who like their daily grind au natural. Instead, you're the type of person who eats, drinks, and sleeps with music. Maybe you still have a ticket stub from a Grateful Dead concert in your wallet, or own a 1400 CD collection, or your iPod is maxed out and The Cloud is your best friend, or you wake up every morning to Eddie Van Halen's version of "Pretty Woman." You have a subscription to

Pandora, and Spotify, and the people at iTunes and Amazon think of you as a model customer. If this even remotely describes your level of interest in music, then by all means this chapter is for you.

Watts All This About Ohms

Think of it a little like a boat's propulsion system. You don't want to move a big boat with a small propeller or a small boat with a big propeller. Nor would you buy just any engine and gear combination to turn any wheel. For example, you don't turn a 20" diameter propeller with an 8-92 Detroit Diesel, or try to muscle a 60" propeller with a 4 cylinder Westerbeke. Either way you're likely to burn up the motor faster than you can say, "Sergeant Peppers Lonely Hearts Club Band."

The same thing's true for a stereo system. After all, a speaker, or driver, is a little like a propeller. Both are output devices. In addition, just like an engine-propeller match, stereo components have to be balanced in terms of their power and impedance, the latter being a combination of electrical resistance and inductive reactance.

"Compatibility is important," says an audio engineer friend of mine. "If you run an improper speaker load, you can damage your amplifier."

For example, connect a pair of 2 Ohm 50 Watt speakers to a 50 Watt amplifier that is rated for a pair of 8 Ohm speakers (an Ohm is a measure of impedance), and the amplifier will heat up and possibly even fry itself.

It won't be very good for the speakers either, which can be permanently damaged by destructive "clipping," a form of distortion caused when the amplifier is overloaded or short on input voltage, a concern in many DC applications.

There's simply no arguing the fact that it's easier and often safer to buy a known system and have it installed by a knowledgeable technician. But some people just have to go their own way, and it doesn;t take a genius to get great sound.

For instance, a manufacturer might list a speaker with the following specification: 100 Watts RMS @ 4 Ohms. In this instance, RMS stands for Root Mean Square, and, continuing with the marine propulsion analogy, it's a speaker's cruising speed. What this means is that the speaker can be safely connected to a 100 Watt RMS per channel @ 4 Ohms amplifier.

However, another manufacturer may use a different rating for the speaker, as in: 100 Watts Maximum @ 4 Ohms. This is a power rating that basically states the speaker can handle a short duration, high-intensity burst of music of 100 Watts. To get a safe approximate RMS or continuous duty rating for this speaker, and it's just a general guideline, you divide by two, which gives you 50 Watts RMS. Hooking up a pair of these speakers to the same amplifier above will give you substandard performance and probably cause premature death for both the speakers and the amplifier. What you need for this 100 Watt Max pair is a 50 Watt RMS per channel @ 4 Ohm amplifier.

Incidentally, there are other ways manufacturers play with power ratings. For example, you might find a power amplifier that advertises 500 Watts. When you look closer, it turns out to be a four channel stereo, and the rating listed is a maximum power rating. Divide the 500 Watts by 2 and you get 250 Watts RMS, then divide by 4 and you come up with 62.5 Watts RMS per channel. Suddenly no so impressive.

Manufacturers also use Peak Power, Full Power, Maximum Output, Instantaneous Peak Power, and Peak Music Power Output, for rating their equipment. These, as well as Maximum Power, are pretty much the same. The only rating that really matters is the RMS rating.

Lastly, many audiophiles will tell you it's all right and even desirable to have a power amplifier that is ten to twenty percent over the RMS rated power of your speakers. This 10 to 20 percent guideline means that a 100 Watt RMS woofer can be safely plugged into a 120 Watt RMS per channel amplifier, provided of course that the impedance ratings are matched. Another thing to bear in mind is that most stereos perform best at a volume setting of 1/2 or less, with the gain on the power amplifier, if one has been added, set at less than 60%.

How Much Rock & Roll?

Now that it's agreed everything has to be matched and balanced with respect to power and impedance, it's time to look at some speakers.

"We always start with the speakers," says Scott Morris, of Ocean Alexander Marine Service, in Seattle, Washington. "Will they be flush mount or box type? How well will they fit the area? Where will the wiring go? Does the customer want to drown out the sound of the engine, or is this just music for dockside?"

Another question you might ask yourself is: How discerning or sensitive are your ears?

"I installed a system for a concert violinist that cost thousands of dollars," says Morris. "I've also installed off-the-rack car stereo systems for a few hundred. If you don't have the range of hearing, you can't tell the difference between the two."

You might also want to keep in mind:

(1) Deep bass sounds are of the long wavelength variety. They require the most power to reproduce because they need a comparatively big driver (woofer) with a long stroke.

(2) Not all music is heavy on bass. If your interest is more in the big band or folk area, you may not need power hungry woofers to get the sound you want. Small two-way or three-way speakers, or a pair of mid-range drivers may suffice.

(3) A general rule of thumb is that to double the sound of what you're currently listening to, you have to multiply the number of Watts by ten. In other words, if you're currently listening to a 10 Watt RMS per channel

marine stereo matched to a pair of 10 Watt RMS marine speakers, and you want to double the loudness, you'll need a 100 Watt RMS per channel system.

For a small boat, however, a 100 Watt RMS per channel stereo is getting up there in power. Here's why:

To find out the power drain on the boat's electrical charging system, double the total RMS value to take into account peak power bursts, then divide by the voltage put out by the alternator to get the current draw. In this case, 400 Watts total peak power divided by 13.8 Volts -- a good average charge from most factory installed marine alternators -- results in a current draw of 28.9 Amps. With a standard 45 Amp alternator, and the electrical needs of sounder, radar, GPS, plotter, bilge pumps, etc. to factor in, doubling your loudness may mean having to buy a new higher capacity alternator, although there are other alternatives. For example, you could install an audio capacitor (basically an energy reservoir) between the battery and the power amplifier. If it were me, however, I'd spend the money on a bigger alternator.

Having decided on the amount of power you need, it's time to figure out what type of speaker. You can go with boxed or enclosed speakers, or go with component speakers, whereby you buy your woofers, midrange drivers, and tweeters individually and mount them yourself. This latter method provides the best possible sound reproduction for a given application but is also considerably more complicated, particularly when it comes to the woofers.

Woofers, Midrange Drivers, TWeeters, and Whizzers

Unlike tweeters and small midrange drivers, which can be mounted in a cabinet or against a bulkhead without concern for altering their performance (we're not talking placement with respect to each other or the listener here, just the actual mounting), woofers have special mounting needs. As the woofer generates sound, it pushes air out its front and back...which introduces a problems for small boat installations.

The sound coming out the back of a woofer is 180 degrees out of phase with the sound coming out the front of the woofer. Consequently, if the woofer is mounted incorrectly, the sounds from the back and the front will interfere with each other and possibly cancel each other out. At the very least, improper installation will cause you to lose much of your low frequency performance. To solve this problem on a small work boat, where cabinet space and bulkhead depth is either non existent or at a premium, you'll need to build a custom speaker box for the woofer, and that brings up an interesting conundrum. Why not just buy a good marine boxed speaker of the two-way or three-way variety?

"In most of the boats we're doing now," says Morris, " we're using two-way and sometimes three-way speakers."

A two-way speaker is one that has a separate tweeter and woofer mounted in the same box, sometimes one over the other to save space. In fact, your typical two-way car speaker has a tweeter mounted either on a post (center mount) or bridge (perimeter mount) directly over the woofer. There's also the dual-cone design, a speaker that consists of two cones, a main cone for lower frequency sound, and a "whizzer' cone for the high frequency sound.

A three-way speaker and a four-way speaker round out the category. The former is a two-way speaker with the addition of a midrange driver, the latter a three-way speaker with the addition of an even smaller tweeter. On a boat, however, the"super tweeter," as it's sometimes called, will do little more than annoy the dog, if he can even hear it.

All outdoor and/or marine speakers are made to withstand the elements. Cones are typically constructed of Polypropylene, while the tweeters are made of Mylar or a corrosion resistant metal, e.g. titanium. Wire and wire connections are coated for water and corrosion resistance, and enclosures or boxes are stainless, marine grade aluminum, or UV stabilized plastic. That's if you buy an off-the-shelf marine speaker.

If you buy component drivers on your own, you can build the boxes and even hard wire everything just the way you want it. Be sure you've done your research, though. There's a lot more to it than what's been discussed here. You've still got to consider output from the head unit to the amplifier, impedance matching if you want multiple volume controls at multiple locations, antennas, speaker placement, and much more.

"Entertainment systems are very much an individual thing," says Morris. "There's no perfect system. What's important is finding one that meets your needs and sound good to you.

###

Theater Sound and More

A long time ago, not in a galaxy far, far away...entertainment aboard a boat was little more than a dog-eared copy of a Zane Gray novel. But, as the showbiz saying goes, "We've come a long way, Baby!" Nowadays—in addition to the smell of the ocean, the cry of the seagull, and the whine of the engine—we have Charlie Daniels blaring on deck at 90 decibels, and reruns of Charlie's Angels in Spanish being piped into the wheelhouse via satellite.

Not every boat owner can or will choose to install the latest and greatest in on board entertainment; obviously, navigation, communications, fish finding, and collision avoidance electronics are first priority. However, for those who want to go to sea with more than a copy of this book, here's a general overview of what's out there, including some advice on what's hot and what's not, and a brief look at the marine versus home entertainment electronics.

AC/DC Stereos, Compact Discs, Flash Drives, and TVs

Small boats without generators or inverters have no choice but to buy 12 volt stereos, CD/DVD players and TVs. But that's OK, because manufacturers have produced some excellent low voltage equipment. However, if I had the money and power (in terms of storage batteries and high output alternators), I'd invest in an inverter and wire my boat for 110 volt service. The electronics choices improve dramatically when you do this, which is to say, you'll have more products and hardware configurations from which to choose. You'll also gain all the benefits that go along with having AC power, like being able to use 110 Volt power tools. Truth is, the money you'll spend on a decent 32" AC/DC color television (around $600) can cover roughly half the cost of a powerful marine inverter. So, bottom line: If you want to watch color television, think about buying an inverter and shanghaiing one of the TVs from home.

OK, let's say you don't want an inverter but still want a TV? No problem. Get a good mobile plan and take your iPad to to sea. If you'd rather go traditional with a TV, you can buy an 15" Jensen Marine flat screen for about $250, a little less than you would spend on a 32" AC flat screen. See what I mean, that inverter is sounding pretty good right now. Anyway, add a digital device like a Chromecast dongle or a Roku box or Apple TV, or by a laptop with an HDMI port, and you're good to go. Still need that mobile plan, though, or a global satellite connection. Pretty slim chance of getting any kind of coverage when you get more than a few miles from shore.

I remember reading back in the 90s that Shaquille O'Neil had a $30,000 stereo system in his Chevy Suburban—and that was not even top of the line! As it turned out, the world champion car stereo belonged to a guy named Earl Zausner of Scottsdale, AZ. His cherry red BMW 540 housed a $46,000 sound system that experts claimed was the very best on the planet. No kidding! They have international competitions for these things, and Earl had ranked as the Big Kahoona several years in a row.

Does this mean you have to spend thousands of dollars on a stereo to get good sound? Of course not. Truth is, the subtle audio qualities of a top of the line system would be lost on the deck of a lot of boats, especially work boats. With a vessel's inherent vibration and noise—winches, engines, machinery, wind and seas—you can't expect to have a discerning ear. This means that maybe you can skip the super high end stuff and go with a reasonably priced car or marine stereo. A 30 Watt per channel USB/SDCard/MP3 player with four speakers. Pretty simple.

Which brings up an important point. What exactly is the difference between a "marine" stereo and an "automotive" stereo. I've had the same $70 Sparkomatic AM/FM cassette deck and indoor/outdoor speakers in my boat for 8 years, and neither the stereo nor the speakers have skipped a beat in all that time. So why bother buying something with the word ' marine' in the name? Obviously, what works in the car or home will work fine on a boat, particularly if it's kept in a dry place.

"Marine stereos have the same guts as car stereos. It's just the housing that's different," says an audiophile friend. "And both still have to be vented.

Boat Tech

Marine stereo manufacturers might not agree. They'll tell you marine stereos are given an extra dose of marinization. In addition to having corrosive resistant plastic housings instead of metal ones, the insides are treated. They'll say they use plastic coatings and silicone on the circuit boards. They'll also use stainless steel connectors and shrink wrap insulators. Stereos can't be 100% waterproof, but they can be moisture-proof. In some boats, they have to be able to handle spray.

For a lobsterman or crabber, or any type of fisherman who works non-stop with wet hands and gloves at a hauling station—as opposed to a draggerman or yachtsman who finds himself in the wheelhouse most of the time—handling the spray is the name of the game. While no company can say for sure that a given stereo will withstand a given set of conditions, it's nice to know you can send a unit back to the factory when something goes wrong.

"A car stereo might have a 1 year warranty," explains my audiophile friend. "A marine stereo might have a 2 year warranty on everything and a 3 year customer protection policy. That policy will set a maximum amount that the customer will pay in the 3rd, 4th and 5th years. And if the company can't fix the unit, they'll give you a new one." Better read the fine print, though.

Finally, before we leave this category altogether, there's one important point to make about compact disk players. Just a few years ago, there was a huge difference in CD players. When you went shopping, you had to be careful because you might buy one for the boat that didn't have enough skip protection. Not so anymore. These days, CD players for cars have been sufficiently hardened for use aboard four wheel drive sport vehicles. If they can handle those they can handle vibration on a boat. That said, with the advent of iPods, flash cards, and other mobile devices, CD players have really become a thing of the past.

TV at Sea

Once the purview of only the largest yachts and ships, satellite HDTV is now available for boats as small as 20', that is if they have a generator, and they're owned by people who want to clunk down about $3,000 for the equipment, which is mostly in the form of a dish antenna and antenna controller. All the rest of the required equipment is exactly what you use at home. In fact, you can take your receiver-decoder box and TV from home and hook it up to the antenna on the boat, as long as you're subscribed to the same service and pay the additional location/receiver fee.

A boat's dish antenna is quite a bit different than a stationary antenna at home, because the boat's dish has to handle pitch, roll and yaw, which it does with an electromechanical pedestal and a combination of gyroscopic and signal strength sensors. While each company employs slightly different methods of stabilization, all track and hold the signal via determinations of proper azimuth and elevation. Does this mean you can watch reruns of the Jeffersons and F-Troop at sea with the same guaranteed reception you would have if you were in the middle of Indiana? Well, it depends. On the boat, the seas, the weather. On a good day, a 70' yacht in the Atlantic, with an 18" dish, can track a Ku-band satellite in 15' seas. However—and this is an important point—there's a huge difference between a 15' post-hurricane swell and a sea brought on by a 40-knot northwester.

Before I go any further, let's get a few acronyms out of the way:

DTH stands for Direct to Home and it is what the FCC uses to categorize satellite to Earth communications and broadcasting industries.

IRD (Integrated Receiver Decoder) is the receiver unit that goes between your television and the dish; you need at least one of these because each network has its own encoded broadcast.

DSS (Digital Satellite System) and DISH are two types of hardware packages, the former a registered trademark of the DIRECTV. the latter of Echostar and the Dish Network.

DBS is short for Digital Broadcast Satellites.

Dual LNB has two coax connections. You can operate up to two satellite television receivers with a dual LNB.

DVB or Digital Video Broadcast, is the broadcast standard for digital radio and television, using MPEG II and MPEG 4 compression. DVB is being supported by all European manufacturers and broadcasters.

Ku-band is the frequency range between 11 and 14GHz; it is often used by communications satellites, carrying the broadcast signal from the satellite to your antenna. It is slowly replacing the older C-band.

L-Band is the frequency range from 0.5 to 1.5GHz. The is the frequency that allows the broadcast to be carried over standard coax cable. The Ku band signal is converted at the antenna and delivered to the decoder via a downconverter.

"DIRECTV is the largest provider in North and South America," says Jim Dodez (pronounced, Dod-ay), Vice President of Marketing for Marine Sales at KVH, Industries in Rhode Island. In fact, a fisherman can access DIRECTV's 175 channels while running down the west coast of the U.S., Mexico, and Central and South America. There's also pretty good coverage in the Gulf of Mexico from Texas to the Carib. What if you don't want DIRECTV? No problem. "If you have Echostar at home and want to stick with it," says Dodez, " all you do is point the dish in another direction and use a different decoder."

Obviously, a vessel's ability to track and hold a signal depends on the vessel's size and location and the type of antenna it has. Satellites are aimed at the continents—where the people are—not at the sea. For example, if you're in Alaskan waters trying to track a Ku-band DBS with an 18" antenna dish, your chance of success is not very good."

It's for this very reason that U.S. based DBS providers like DIRECTV and Echostar/DISH advertise that their broadcasts will extend no more than 100 miles beyond the borders; they actually go farther, but not far enough to reach the waters of Alaska or the 200 mile limit. This means that vessels in the Gulf of Alaska, Carib, Hawaii, and/or mid-oceans, need what Tim Allen of Home Improvement refers to as...more power. Offshore coverage is a function of antenna size. In other words, when it comes to uninterrupted global satellite coverage, "size does matter," and antenna's can be custom manufactured up to about 9' in diameter.

What's the future hold? According to recent reports, the DTH/DBS industry is moving toward Ka-band satellites because of that frequency range's ability to carry more bandwidth.

Theater Sound

If you're going to spend the money on a DBS system and the theater experience, you might as well listen to it the way it was meant to be heard...in Dolby Surround Digital.

The original Dolby Surround Sound was introduced in the seventies with movies like Stanley Kubrik's, A Clockwork Orange. The system used multiple channels (right, left and surround) and a wall of speakers to fill the theater with sound. Unfortunately, the front and back sound interfered slightly with the dialogue. Consequently, in the late 1980s, Dolby introduced Pro-Logic. What

Pro-Logic does is improve signal separation and add a fourth channel, a center channel, to better accentuate the dialogue; in the majority of movies, as much as 70% of the sound comes through the center speaker.

Dolby AC-3, a.k.a. Dolby Stereo Digital (the industry name) and Dolby Surround Digital (the consumer equivalent), was brought to the screen for the very first time in the 1992 release of Batman Returns. What Dolby Digital does is add one ' real' channel and one 'phantom' channel. It turns Pro-Logic's monophonic rear channel into a right and left channel—providing true stereo and higher power in the process channel. This makes a total of 5 and 1/2 channels, or as you will most likely see it written: 5.1 channels.

Now, to further complicate matters, F/X wizard extraordinaire, George Lucas. of Star Wars fame, decided that the huge theater sound of Dolby Pro-Logic didn't translate that well to the home. In an effort to correct this problem, he had his chief engineer, Tomlinson Holman, invent a technique for digitally reforming the sound, which in big theaters is partially absorbed by the seats and the viewers sitting in them. According to Lucas' company, THX certification provides a better acoustic match to the room the movie is being played in by Re-equalization, Timbre Matching, and De-correlation of the prerecorded Pro-Logic signal. On other words, THX is not a recording process or a film sound format. It's a post-production quality assurance enhancement that improves the sound criteria. If a piece of equipment or audio production carries the THX label, it means the device or production got the

THX seal of approval from George Lucas's Studios.

At present, Dolby Digital 5.1 is the accepted standard for DVD. Dolby Digital Plus 7.1 is the accepted standard for BlueRay, HDTV and DTH TV. And Dolby TrueHD 8 is the up and coming standard for BlueRay, HDTV, WebContent HD, and HDDBS.

NOAA GRAPHIC OF SIDE SCAN

CHAPTER EIGHT
THE SOUNDER

Fishing for a Fish Finder - SONAR
Given the incredibly competitive nature of fishing, it's easy to understand why the owners of multi-million dollar fishing operations equip their vessels with scanning sonars. In fact, more than a few boats have installed two and even three different models—a financial commitment that rivals and sometimes exceeds the cost of the vessel's main engine(s).

And yet, scanning sonars are not just for the largest of factory trawlers and mid-ocean tuna seiners. Almost every boat owner who hunts for fish can benefit from the technology. True, some equipment is prohibitively expensive and cumbersome for small boats, but if you're in the market, you can find a model that fits your boat and budget.

What do you look for when it's time to start shopping? It's a complex issue, but here's a primer that can help get you started.

According to Gene Hill, who invented the omni-directional sonar back in 1969, and is currently President of MAQ Sonar in Lunenberg, Ontario, the first thing to do is understand the type of work the vessel does. Will it be working in rough seas, calm seas, shallow water or deep water? And what are the target species? Will they be midwater fish, bottom fish, or both?"

These questions are important because you need to not only match the equipment to the boat but also the intended use. Like everything else in life, there are trade offs. The first one has to do with range. If you need to see more than a 1/4 mile, you have to go with a lower frequency sonar. The lower the frequency the greater the range. But if you want to see smaller targets, you need the higher frequencies. Typically, for smaller targets, you would use a higher frequency, narrower beam sonar.

Need some examples?

"Normally you see mackerel at 180 kHz," says Hill, speaking with regard to the frequencies used by MAQ sonars. "The fish are not as strong at 90 kHz, weak at 45 kHz, and very week at 22 kHz."

Hill adds another wrinkle to the equation. He says his customers in Korea fish for anchovies with 180 kHz machines, but in Spain they fish for them using 90 kHz machines. Are water temperature, salinity, and other localized sea conditions that critical?

"Yes," says Dave Quarders, Sales Manager for JRC North America, in Seattle, Wash. "The warmer the water, the greater the distance the beam can travel. In cold water, sometimes you have very severe thermoclines that can reduce the performance. So yes, it can make a big difference."

In some cases, prop wash is a factor. "The tuna fleets," says Hill, "are using high frequency sonars to see the fish closer to the wake. Ordinarily, you can't see through the wake of a boat, because of all the bubbles, but the higher frequencies can travel between bubbles of 1/4" diameter."

Bubbles are one thing. Turbulence in the form of wave action is another. "If you have a rough bottom or the seas are choppy, you'll get echo," says Dennis Soderberg, Vice President of Western Marine Electronics (WESMAR), in Woodinville, Washington.

What do you do to avoid interference from bottom or surface echo? You choose the right beam width.

Beam width is the second important consideration. While it's advantageous to cover the largest area possible in a given sweep, a wider beam is not always the ideal transmission characteristic. When working in shallow water or searching for targets that live on the bottom or close to the surface, a narrower beam may be the smarter choice.

Picture a conical beam of sound leaving your boat and heading out at zero degrees relative. The farther the beam travels, the wider it gets. At some point, it will hit the bottom (or the surface) and bounce back. When this happens, the return from the bottom (or the surface) interferes with the return coming from targets inside the beam.

"The bottom is always a better echo," says Soderberg, "which means the bottom return will cover the echo for the fish."

How far away can you spot fish if you're in shallow water?

"With MAQ sonars," says Hill, "which have the narrowest vertical beam of any of the Omni-directional sonars, the range is limited to twenty times the depth. If you're in 100' of water, the maximum range would be about 2,000'. But this varies with the type of bottom; a muddy bottom would give you more range, a hard bottom would give you less. In deep water, we can achieve ranges of 10,000 feet or roughly two miles."

Is there a magic combination of beam width to frequency for the type of fishing you do? Not really. But there are several that will work better than others. Fortunately, manufacturers provide a wide selection of beam widths and frequencies.

What's Out There?

Not everyone needs or can afford the very best in sonar technology. Truth is, costs can run in the hundreds of thousands of dollars, and equipment can weigh several hundred pounds. A small seiner trying to distinguish between a school of herring and a school of mackerel, and a large factory trawler trying to pick out a school of pollock, won't be using the same equipment, or be required to spend the same amount of money for satisfactory results.

Given this fact, how do fishermen get the best bang for their buck? For one thing, it's important to know the different kinds of sonar technologies available.

Omni-directional sonars are the most expensive of the three. They use multiple transducers to send out multiple beams at the same time. The actual "scanning" is done electronically by a processor, so there are no moving parts in the transducer assembly.

Searchlight sonars use one transducer that is moved around mechanically, hence the name...searchlight. They have an assembly that allows you to point the beam in any direction horizontally and vertically.

Finally, hybrid sonars combine the characteristics of both searchlight and omni-directional technologies.

A USED WEBASTO FURNACE GOING INTO MY BOAT

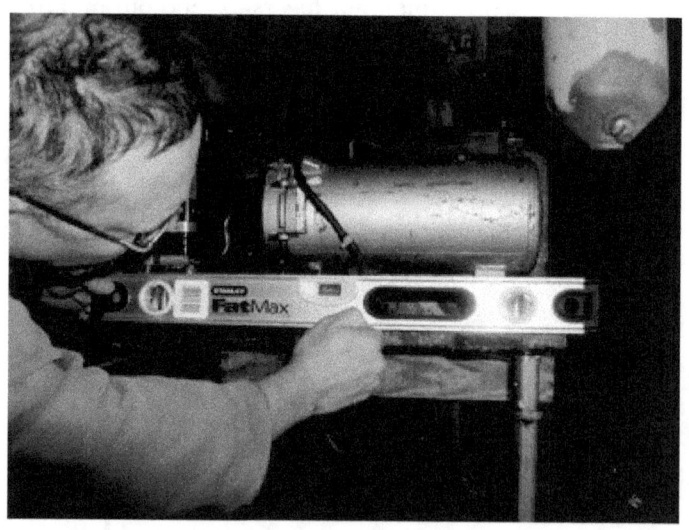

CHAPTER NINE
FURNACES FOR BOATS

Snug as a Bug

Coming off a cold deck into the luxurious warmth of a heated cabin makes a tired body feel good. Problem is, on small boats, you're either huddled around a stove that somebody's drying their socks on, or revving an expensive engine for no reason other than to feed more coolant to a bus heater.

That's a good case scenario. A bad case scenario is that you're picking black snot out of your nose every morning because the old 'black booger' burner puts out more carbon than a steam locomotive. Or you're sleeping with your arms around a fire extinguisher because, when the wind blows, the burner sounds like an M-80 going off in a 55 gallon drum. Or maybe you're dumb enough to be working with a flue-less heater designed solely for outdoor use, in which case you're dealing with all of the above, plus you're drenched in condensation.

"You kind of put the threat of being soaked, blown up, or asphyxiated out of your mind," says a commercial fisherman I know who lives and used to fish out of Spruce Head, Maine. "Sure, you can die, but at least you're warm."

Indeed, short of setting it aflame, there isn't much I haven't tried to do to heat a boat, which is why, after all these years, I've finally bitten the bullet and purchased a marine furnace.

Burner Pot, Engine Driven Heater, or Furnace?

The burner pot stove (Dickenson, Force 10, Refleks, Sigmar, Taylors) has served the industry well. The stoves are rigged and hold up to conditions at sea. But these things are more stove than furnace or boiler, which is to say, their functionality lies in their simplicity.

Burner pot stoves have a reputation among fishermen as being kind of finicky, hence the nickname, the old Black Booger Burner. Draft problems set the flame to burn unevenly, too low, or too high, and this causes incomplete combustion of fuel, which leads to puffing and/or smoking. This also happens when the stove is installed wrong or skewed too far one way or the other. The reason for this is that it's a gravity fed, non-pressurized system, with a very simple metering valve. In order for the unit to run properly, the valve and the burner have to be fairly level with each other. A bad installation, and/or too much pitching or rolling, will either flood the burner or starve it of fuel.

Albert Bunker, a lobsterman who used to fish out of Matinicus Island, Maine, and who has seen some pretty rough weather running pilots to and from ships in Penobscot Bay, never really had this problem. "I got one of those small Refleks stoves," he said to me many years ago. "I turn it on and it goes a month." Although, he cautions: "You can't run it low or it carbons up. I run mine at [a setting of] two and a half or three."

Bunker added that he never really had problems, except when coming up alongside a big ship, at which time he might get a puff of wind down the stack. Otherwise, he said, the seas and the wind didn't make much difference.

I think Albert was overly complimentary. My old pot burner sucked fuel like a diesel locomotive and spewed black soot if I neglected to shut it down and clean it out biweekly, which is an important point to remember. It's critical to keep that burner plate clean and smooth as a smelt. The slightest divot or scratch will contribute to an uneven burn, and it's the uneven burn that causes smoke. A new, polished bottom plate, doesn't require cleaning. Also, I find it helps to burn kerosene in lieu of diesel.

On their good side: They look good. They're quiet. They're cheap to operate. They heat without electricity. And, most important, you can cook on them, which means never again having to clean Dinty Moore stew off the engine's valve covers.

On the other hand, automotive heaters, with a heat source supplied by hot coolant from the engine, have their own set of pros and cons. On the upside, they use waste heat from the engine, so there's no lighting them, no flame to worry about, and no beating them in terms of dollars and cents.

On the downside, the heater off the engine won't preheat the boat, as it obviously doesn't work when the engine is off. It also adds a risk factor to the engine's cooling system: an air lock in the circuit, or worse, a blown heater hose, can have costly consequences.

Last on the list are the hot air furnaces and the coolant heaters, a.k.a. hydronic' heaters. Espar, Hurricane, and Webasto make both. Toyoset and Wallas make only the former.

These furnaces are true central heating units for small boats. They can be plumbed or ducted to various parts of the vessel and set to feed off the main fuel tank. (Burner pots are best served by a separate fuel tank).

With coolant heaters, you can actually zone the system. They can be connected to radiators, heat exchangers, blower fans, and be used to preheat the engine. On the plus side, they're compact, made to operate in various sea states, and can put out a lot of heat to different areas of the boat. On the downside, they're noisier than a burner pot, more expensive to buy, need electricity in order to work, and are a lot harder to install.

Installing a Coolant Heater

Given the above, and the fact that I didn't want to spend too much money, I decided to look for a used marine furnace for my boat, a fifty foot Chesapeake deadrise style vessel presently rigged as a party boat but which has seen service over the years as an urchin tender, tub trawler, and gillnetter. As luck would have it, I found a 20 year old Webasto Thermosystems 40,000 BTU coolant heater through a local shop & swap paper called, Uncle Henry's.

After looking over the unit, and contacting the manufacturer, I came up with the following installation notes:

(1) One of the biggest concerns in any stove or furnace installation is making sure carbon monoxide (CO) emissions don't pose a threat to the crew. Venting the furnace out the stern, particularly if the engine's main exhaust is there, makes the most sense. If you're in a room with CO at 12,800 ppm, without a break for fresh air, death can occur in less than 3 minutes.)

(2) You want to keep water from entering the exhaust, and, unless you're regularly backing hard into a sea, an outlet high in the transom poses little chance for this to happen.

(3) The shorter the exhaust run, the less back pressure, the better the furnace will run. In fact, Webasto furnace specifications call for an exhaust run of less than 10', with the total number of bends in the pipe not to exceed

270 degrees.

(4) Exhaust outlets must be insulated.

(5) The exhaust outlet must terminate in a zero pressure area. Plumbing it into another stack or exhaust is not recommended.

(6) It is advisable to include a water trap and/or goose neck into the exhaust system if it exits through the hull.

(7) In positioning the furnace, bear in mind the expansion tank must be higher than the furnace, while the coolant circulation pump must be lower.

###

Manufacturers List
Burner Pot Stoves and Heaters

Dickinson Marine (1997) Ltd.
#407 - 204 Cayer Street
Coquitlam, British Columbia
Canada, V3K 5B1
1-800-659-9768

(Taylor Stoves)
Blakes Lavac Taylors Co,
13 Harvey Crescent
ÊWarsash , Southampton
United KingdomÊ
SO31 9TA
00 44 (0)1489 580580

Force 10 Marine
23080 Hamilton Road
Richmond BC
Canada
V6V 1C9
Toll-free 1-800-663-8515
Tel: (604) 522-0233

Refleks Olieovne A-S
L¿rupvej 17
DK-5750 Ringe
Tel.: +45 62 67 12 68

(American Distributor)
Hamilton Marine
PO Box 227
155 East Main Street
Searsport, ME 04974
Maine, USA
(207) 548 6302
(800) 548 6352

SIGMARINE LTD.
23080 Hamilton Road
Richmond, B.C.
Canada
V6V 1C9
604-737-0101

Coolant Heaters and Forced Air Heaters

Hurricane Heating Systems

Robert Bernstein

International Thermal Research Ltd.

Vancouver, WA 98661
Telephone 360-993-4877 Toll Free 1-800-993-4402

ESPAR Heating Systems

(West Coast Distributor)
Boat Electric Company
2520 Westlake Avenue North
Seattle, WA 98109-2234
(206) 281-7570

(West Coast Distributor)
Ocean Options
95 Riverside Dr.
Tiverton, RI 02878
(401) 624-7334

Webasto Thermosystems Inc.
North American Headquarters
3333 John Conley Drive
Lapeer, Michigan 48446
1800heater-1

(Webasto Dealer Contacted for this Article)
Great Lakes Marine Specialties
4133 West 45th Street, Minneapolis, MN 55424-1040
800-821-0207 or 952-925-1285

Forced Air Heaters Only

Toyoset Heaters
Toyotomi, U.S.A., Inc.
604 Federal Road
Brookfield, CT 06804
tel. (203) 775-1909

Wallas-Marin Oy
Karrykatu 4
20780 Kaarina, Finland
tel. +358 (0) 2 412 0500

NORTHERN LIGHTS 944T

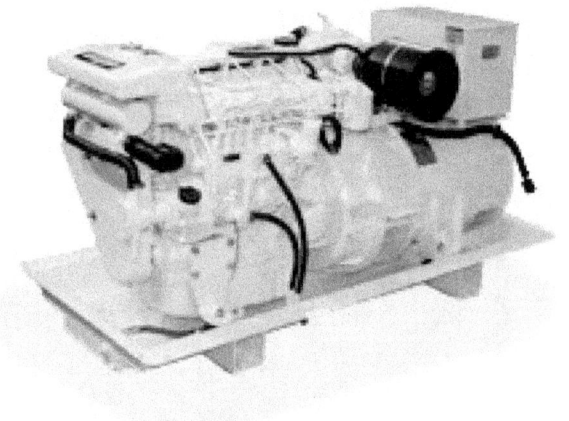

CHAPTER TEN
GENERATORS

A Primer

Generators don't just provide the creature comforts of home, they add security and peace of mind. Whether they're charging the batteries, or a compressor and air-storage tank, for the main engine, lighting the decks and superstructure to be better seen at night by other vessels in the area, or affecting the ability to make repairs by supplying electricity to power tools, they significantly increase a vessel's overall strength and endurance.

Most big boats—and in this case "big" refers to a vessel with high capacity AC appliances and other substantial electrical demands—require at least two generators to maintain proper service. If ship's power is derived from one unit, two are installed so that a routine operation and maintenance schedule can be kept. Usually, one generator will run for a week or two while the other is serviced and allowed to "rest."

Robert Bernstein

What Makes Them Tick?

No matter what kind of generator you have—rotating armature-type, or stationary armature-type, also called a rotating field generator—the principles of operation are the same. Essentially, mechanical energy is converted to electrical energy by moving a coil through a magnetic field.

In all cases, windings of copper wire and various electrical components (rectifiers, regulators, capacitors, diodes, etc.) are combined in a system that consists chiefly of two main parts: a rotating assembly called the rotor, and a stationary assembly called the stator. Just as you would expect, the output coils or armature—the part of the machine that carries the current—is located on the rotor on a rotating armature type generator, and on a stationary armature design, it's on the stator. In the first, the magnetic field is stationary and the coils that induce the voltage rotate. Most of your DC generators are of this type, as it's impossible to get DC from a stator.

In the rotating field generator, the coils are on the stator and the field rotates. Most of your AC generators are of this type.

Boat Tech

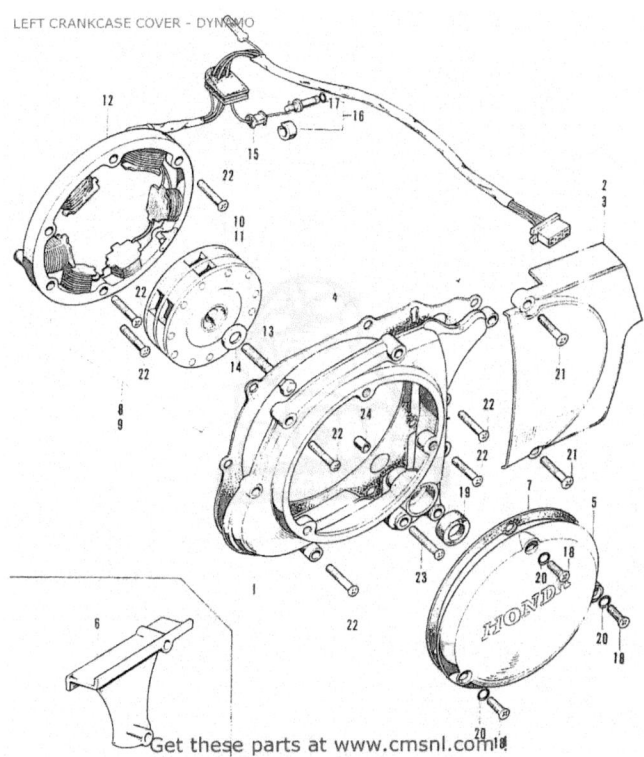

Rotating Armature Type

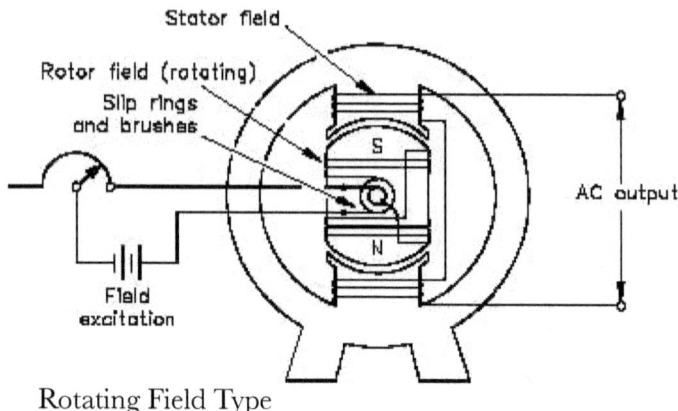

Rotating Field Type

Also, while a few field windings on smaller generators are self-excited—i.e. they retain a certain amount of magnetism after shut-down—the majority on larger generators need to be "kick-started" with direct current. This is done with an external source like a battery, and/or with an internal exciter circuit. These circuits are separate pairs of stator/rotor windings or permanent magnets in the stator/rotor assembly, sometimes both. In some cases, a third "pilot" exciter—usual permanent magnets—kick starts the main exciter. One way to look at it would be as a generator within a generator within a generator.

The other parts of the generator include: (1) the output control circuit, which feeds the vessel's distribution panel; (2) sometimes a set of windings for battery charging, and (3) the brushes and slip rings in brush-type generators.

It should be noted here that brushless generators are

probably more popular than brush-type generators, especially in smaller sizes. The additional number of moving parts exposed to the effects of friction in a brush-type generator require more maintenance. Today, there are many more brushless generators than brush-type. On the other hand, bushless generators are more expensive and more complicated, usually with more costly repairs when they do go down.

Keeping Them Going

According to Steve Davis, a former production control specialists with the Baylor Generator Motor Group in Sugarland, Texas—original equipment manufacturers and service providers for a broad range of generators—a well maintained generator can last a very long time. "We have customers with Delcos that date back to the mid-thirties," he says.

Davis and others who work on generators recommend the following: Make sure the windings are clean and not saturated with oil or grease; Routinely check the brushes in brush-type generators; Check the oil in the bearings (in units that aren't permanently sealed); And check the resistances in all electrical components.

"They have to be protected from moisture and grease," says Davis. "Dirt causes heat, and heat breaks down insulation. Also, keep all air passages open. They're no different than the valves going to your heart."

John Warner, a salesman for Eastern Electric in Seattle, Washington, agrees. "If equipment gets dirty and especially if diesel fuel gets into the generator, you're looking at reduced longevity."

Of course, there are times when generators need a little more than just routine attention.

"Depending on size or access, the best thing is to take the generator-end off and ship it to the shop for servicing," says Warner. "If you can't do that, because the cost is prohibitive, for example, you have to cut through the hull to get it out, then your only alternative is to clean it in place."

In or out of the boat, the best way to clean the generator is to disassemble it and use a commercial steam-type cleaner that actually sprays the unit with soap and water. This, in combination with testing individual components and replacing parts where necessary, is the usual procedure. But drying takes time, which may not be available. In a situation where time is a factor, a technician may opt for a combination wipe down/ resistance check in lieu of the more thorough steam cleaning.

The wipe down is exactly what it sounds like. Components are disassembled and wiped with a rag, and the unit is kept relatively dry so that it can be reassembled and started without too much delay.

Rebuilding a Generator

When a resistance check of a generator's components yields values that exceed the manufacturers specifications, it's time for the thing to be rewound or replaced. "If a major component goes, like the main exciter rotor or the main stator," says Davis, "the generator has usually gone beyond the point where it can be rebuilt."

However, until that point, rebuilding a generator is a perfectly viable option, as long as you find someone with the proper qualifications to do the job.

"The main thing you want to look for is a history of repair," says Davis. "How long they've been doing it? What equipment do they service?"

Insulation is critical, especially for a generator on a boat. For example, in the case of Baylor AC generators, rectangular copper wire is insulated with an enamel/mica composite and formed into the stator coils. The coils are then insulated with mica-glass tape, mounted, and further insulated in a VPI process. "VPI stands for Vacuum Pressure Impregnated," says Davis. "It's an insulation system that keeps foreign matter out by fully encapsulating the unit in epoxy resin. There are other methods, like Glyptol, but VPI is the best."

The other insulation used in a generator includes the separators between the coils and the core, the varnish used at the turns and elsewhere, and the epoxy used in the VPI and the wet-winding process. Wet winding is an additional layer of insulation.

Of course, choosing to rebuild is not always the easiest thing to do. "Rewind costs make a rebuild an attractive alternative when compared to buying a new generator," says Davis. "And any good rewind shop can have one done in three days. But there's always the possibility of fixing one component and having another one breakdown later. I have to say that the decision depends on the situation...how old the machine is, and where and how it's being used."

This is generally the bottom line. Was the failure the result of one or two components, or was it caused by more than that? The decision to rebuild should be based strictly on the overall condition of the machine. For example, why spend money on a rebuild if the frame is rusted out?

TRICO'S OIL SAMPLE VACCUM PUMP

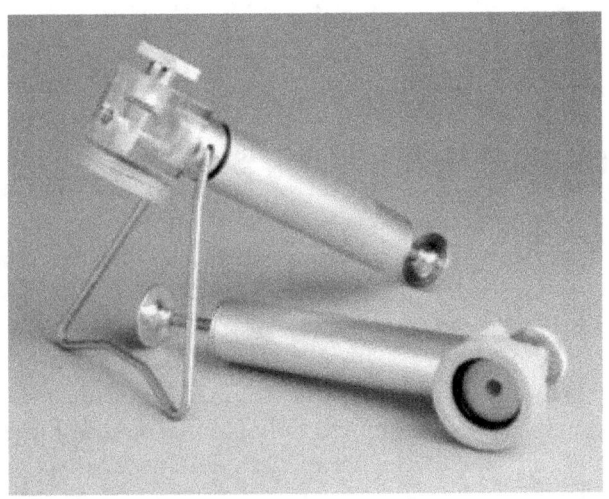

CHAPTER ELEVEN
LUBE OILS

Condition Based Maintenance

In a perfect world, we would all opt for a program of pro-active maintenance. We would take routine oil samples in a prescribed manner, send the samples to the lab, and learn how to interpret the results. We would squeeze the most life out of our lubricants and at the same time make sure the engine never operated with contaminated oil. In other words, all our oil changes would be determined by the results of oil analysis and not on a predetermined timetable.

Why? Because of all the engine diagnostic tools and technologies—vibration analysis, thermography, etc.—oil analysis is the least expensive and most effective way to avoid equipment down time.

But what happens when I take samples from the same engine and send them to several different labs? Do all labs provide the same service? Will they all report the same results?

Let's find out.

Getting Started

Samples were sent to five companies in all,

Oil Analyzers Inc.
Blackstone Laboratories Inc.
Herguth Laboratories, Inc.
PdMA Corporation
Texaco (Texscan)

I found out about Oil Analyzers Inc. and Texaco locally, from my local Amsoil dealer and NAPA Auto Parts, respectively. Herguth, PdMA, and Blackstone were initially contacted on the Web. A search of the words "Oil Analysis" brought their web pages up and an e-mail to the individual companies yielded a mail packet in less than a week. A fourth company, CAC, of Eatontown N.J, also responded via the Internet, but at $120.00 per sample, their oil analysis fees were three to 10 times more than the other companies.

[NOTE: This test was done many years ago and prices have no doubt changed. Please contact companies directly for updated info.]

Blackstone Labs sent out a sampling kit immediately after receiving my request over the Internet. It was the first to arrive in the mail. The Texaco kit was picked up at NAPA. and the kits from the other companies were ordered over the telephone.

At Herguth and PdMA, they gave me the choice of using their sample bottles or my own. To save time, I said I would find my own. However, in retrospect, I should have just accepted theirs. As it turned out, two to three ounce plastic bottles with tight fitting caps were not that easy to find. I struck out at Walmart, two auto supply stores, and one chandlery. Eventually, I found what I needed at a store that sold camping equipment; it seems backpackers use a lot of little plastic bottles. Meanwhile, the Oil Analyzers Inc. kit took about 10 days to arrive due to a back order delay.

Oil Analyzers Inc. and Texaco were the only sample kits for which I paid in advance. The others required that I enclose a check in the envelope with the sample. Of course, all the companies would have set me up with a charge or billing account had I planned to start a program of routine testing.

Finally, written sampling instructions were pretty much the same i.e., warm engine, avoid contamination, etc. However, instructions over the phone were less detailed and less stringent. In fact, Herguth told me it would be all right to draw the oil from the sump, as long as I didn't take it off the bottom. More about this below.

Recommended Sampling Methods

According to James C. Fitch, the creator of the pro-active maintenance approach and the founder of Diagnetics, Inc. in Tulsa, Oklahoma—a leading provider of oil analysis training and technology—the person taking the sample shouldn't just do an oil change and skim the recommended two to three ounces of lube off the top. Nor should the sampler pull the drain plug and this is that wear particles and other contaminants settle out/or adhere to pipe and container walls when the fluid is cold and/or inactive.

Ideally, a sample should be taken upstream of the oil filters—while the engine is running and at operating temperature. One way to do this is to tap the feed line to the filter(s) with a three way diverter valve or an assembly that includes a t-fitting, hose and valve. The diverter valve, or combination hose and valve assembly, should be installed so that the sample is drawn from the outlet that is 180 degrees from the inlet, not from the outlet that is at 90 degrees. Why? Because blow-by causes the particles and contaminants to bypass the elbow when the engine is running. In other words, they take the path of least resistance.

The sample should also be as pure as possible. Bottles, valves, hoses, pumps, etc., if not brand new, should be flushed and cleaned before the sample is taken. In fact, for dynamic sampling (engines running) there are special test ports and probe-tube bottle attachments that enable you to take samples without having to remove the bottle's cap. Diagnetics, Inc. and its parent company, Entek IRD of Milford, Ohio, specialize in the sale and distribution of this type of equipment and service.

For static sampling (engines not running) there are vacuum pumps that do the same thing i.e., provide a clean interface with the sample bottle. The PdMA Corporation in Tampa, Florida sells a plastic and metal suction gun that will draw oil with a pressure equivalent of 27" of mercury or about 13.5 psi.

My Sample

The sample engine: A 1978 Cummins VT-903 400 hp marine diesel with 6,000 hours of total running time. The oil: Texaco Ursa Super Plus SAE 40 supplemented with 1 liter of Power-Up synthetic oil additive. Running time on the oil: Approximately 80 hours.

I wasn't planning on starting a routine maintenance program, and even if I was, I wouldn't install a three way diverter valve in my oil pressure line, at least not a permanent one; I have enough things to worry about when I'm underway, I don't need to wonder whether or not the valve is vibrating loose. By the same token, I wouldn't want to add anything to the engine e.g., a 90 degree elbow, that would interfere with the factory-engineered flow of lubricant. However, if I were to start a routine program, I'd make up a t-fitting, hose, and valve assembly that I could attach and remove prior to and after each sampling procedure.

Admittedly, the samples for this article were taken under somewhat less than ideal circumstances. None of the samples were drawn while the engine was running. Oil was taken from the return line at the filter head, not the feed line. And I poured the oil directly into the sample containers, instead of drawing it with a special oil sampling pump.

On the positive side, I discarded my first sample, and I filled all the bottles at roughly the same time. For the record, two labs were given samples that were taken from a cold engine. Two labs were given samples from a hot engine, and one lab was given one sample of each.

Results

All the labs tested were very accommodating and organized. For the most part, there was no waiting, no busy signals, and no voice mail. I gave them my name and/or company name, and they gave me a simple thumbs up or thumbs down on my oil over the phone. They then arranged to FAX me or email me the report, a duplicate of which was also sent by mail.

Obviously, the five companies have their own types of reports. They don't all test for the same things, and they don't all present the data the same way. While I would have liked to show the reports individually, the availability of space dictated otherwise. Therefore, all the data has been combined into one spreadsheet. Bear in mind that wherever there were different designations for the same material (glycol/anti-freeze, insolubles/solids, Total Base Number/TBN), both were listed. This was done so that it would be easy to see which company used which designation. Also, not all the companies go by the same standard of measurement. For example, Herguth's "Abs. Soot" is not the same as Oil Analyzers Inc.'s "Soot".

PdMA had the contaminants, as well as minimum and

low warnings for TBN and Viscosity. The Blackstone lab report included universal averages.

All the companies were fairly consistent in their reporting, except for Oil Analyzers Inc., which had a strangely high value for boron, and a comparatively high value for aluminum. Boron would be considered a potential oil additive, possibly a component of the Power Up Oil Additive I used in my engine, while aluminum is a wear metal. These values were either anomalies in the sample, or they were accurate values that weren't picked up by the other companies. The engineers at Oil Analyzers, Inc. didn't consider the results significant, but they did suggest keeping in them mind as part of a longer term track and trend data collection program.

This last recommendation—track and trend data—is the key to condition-based maintenance. Because while today my engine oil tested satisfactorily, tomorrow or two weeks from now...who knows?

Finally, a recommendation of my own: If you're looking to begin a maintenance program that includes oil analysis, my advice is to pick one lab and stick with it. Because, as James C. Fitch points out in one of his company's publications, "No two labs can perform...with equal performance and excellence.

Participating Oil Analysis Companies
Participating Oil Analysis Companies

PdMA Corporation (813) 621-6463; FAX (813)

620-0206
5909-C Hampton Oaks Parkway
Tampa, FL 33610   Comprehensive Test includes Basics, Wear Particle, Asic or Base.

Blackstone Laboratories
4929 South Lafayette Street
Fort Wayne IN 46806
(219) 744-2380 FAX (219) 745-2200

Condition Analyzing Corporation
Telephone: (732) 542-5588
Fax: (732) 542-2967
Postal address: 23 White Street, Eatontown, New Jersey, USA 07724
Electronic mail: CAC@ConditionAnalyzing.com

Fabick Power Systems
SOS Oil Analysis
101 Fabick Drive
Fenton, MO.

Herguth
101 Corporate Place
Vallejo CA 94590
1-888-HERGUTH (437-4884)

Conam Inspection Inc.
Cleveland Technical Center Division
101 W. Mohave
Phoenix, AZ 85003
(602) 253-6515
FAX: (602) 252-4639

Robert Bernstein

WWII ERA RADAR

CHAPTER TWELVE
RADAR

Myths

There has been a lot of confusion in the past about how radars, single side bands, and satellite communication systems work, and how dangerous their transmissions are. Some people think you can get seriously hurt by standing too close to the antennas when they're transmitting. Others think they're completely harmless. A few people see them in the same way they see microwave ovens or x-ray machines. With so many different viewpoints, we thought it would be a good idea to find out what if any danger really exists.

"It's a vague area," says Ralph Sponar III, a technician who does a lot of work on high power radars —10 kW to 30 kW—for United Radio, Inc., in Baltimore, MD. "Speaking personally, I never get in direct line with the beam."

The Beam

Robert Bernstein

What we're talking about here is a radiating beam of energy. But it's important to note that the word "radiation" is a relatively meaningless term for something that has been figured out by mathematicians. In other words, radiation is the emission and propagation of "imaginary" waves or lines.

Don't try to picture it. Instead, simply keep in mind that (1) radiation is a term for something we can see the results of but don't really understand, and (2) radiation comes in many different forms.

Basically, there are three categories—electromagnetic (EM), sound, and particle—and multiple subcategories.

For EM radiation, going from the highest frequency to the lowest, the subcategories are: gamma rays, x-rays, ultraviolet (UV), visible light, infrared (heat), microwave, and radio.

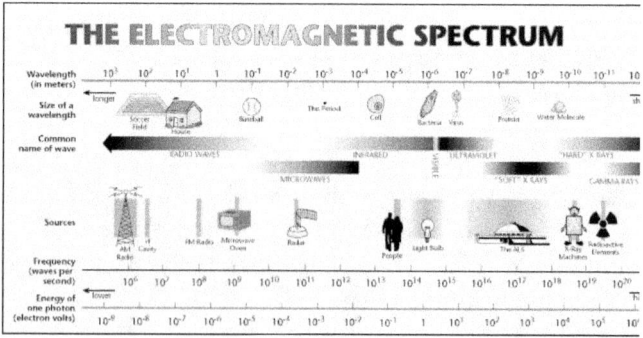

For the sound or acoustic category, the subcategories are infrasonic, sonic, and ultrasonic. And for the particle category, the subcategories are known primarily as alpha and beta emissions, e.g. the decay products from radioactive material.

Which are the worst? Actually, it depends on the strength and type of the radiation and the length of time you're exposed to it. In fact, what scientists refer to as "ionizing" radiation is considered the most harmful. In other words, radiation that "excites" or "modifies" the material it passes through is typically viewed as being the worst. In this regard, alpha and beta emissions, and gamma-rays and x-rays can cause the most harm in the least amount of time. In contrast to this, UV rays, visible light, infrared, or radio waves (such as the microwaves from a radar) are relatively harmless over short time periods, which isn't to say they can;t be fatal if ignored or abused.

Do mariners use all forms of radiation? No, but they use a surprising number of them.

What We Use and Where

Acoustic radiation has applications in sounding machines and sonars. In fact, we use sound waves in much the same way a dolphin uses echo-location. Are they completely safe? On a boat, mostly yes. But the answer really depends entirely on the strength and/or the duration of the signal. For example, the sperm whale uses one type of sound wave to immobilize its main source of food, the giant squid. It's a short burst of a very low frequency wave, and it's deadly. Another type of sound wave, say, the higher frequency whine from a diesel engine, can damage the human ear over a long period of time.

In the very high energy arena, ionizing radiation—X-rays and gamma rays and the particles emitted from radioactive materials—are so dangerous they really have no place on a boat. The two exceptions are found in some types of gas detectors and watch dials. For example, the green luminescence in your timepiece depends on the alpha emissions from a tiny quantity of radium. This proves, of course, that when handled properly, even dangerous radiation can have safe and sometimes—as in the case of x-rays in medical diagnostics—life-saving applications.

Take UV radiation. It's found in water purifiers that are occasionally used with marine fresh water systems and reverse osmosis or evaporator-type watermakers. A UV system might also be found in the processing room of a factory trawler. Helpful? Yes. Damaging? Possibly, especially if you open one of the devices that kill bacteria and other microbes. In fact, looking into a UV bulb would be a lot like looking directly into the sun. Not a good idea.

Meanwhile, of all the forms of radiation, the radio wave has the widest application in the marine industry. It's what we use in GPS, Loran, SSB, VHF, cellular, radars, and satellite communication systems.

Why do We Use Radio Waves?

Radio waves have certain qualities that make them perfect for marine navigation and communication.

(1) They're the least dangerous of all types of radiation.

(2) Depending on their frequency, they can either penetrate or be reflected by a target. In other words, you can receive a radio transmission on a handheld while lying in the bilge of your boat.

(3) They're not easily diffracted or refracted by rain, fog, ice, snow, dust, pollution, etc., not the case with UV light, visible light and infrared radiation.

[This isn't to say that mariners don't use these latter forms of light, because they do. The military uses them in laser guidance, range finding, imaging, and data communications systems, and we're starting to use them as rangefinders in some high-end marine binoculars. However, lasers can be diffracted and refracted, and they can't penetrate materials without causing some type of damage.]

(4) Like all other forms of EM radiation, radio waves travel at the speed of light. And this—combined with the second quality—is what makes them work so well. Take the radar.

Simply put, a radar—an acronym for (Ra)dio (D)etection (a)nd (R)anging—sends out a signal, then receives its echo as it's bounced back from a distant object. Because the direction of the antenna is known at the time of the initial transmission, and because the time between the transmission and the reception of the echo can be measured, the radar can determine the distance and direction of the object in question. You might ask: But how does the antenna receive the echo if the antenna is always moving? In other words, when the transmission is initially propagated, the antenna is in one position, but when the echo comes back, it's in another. How does it have time to "catch" the echo? The answer is simple.

Radio waves and all other forms of EM radiation travel faster than anything known to man. This velocity, in a relative vacuum, is roughly 186,283 miles per second. At this rate, a radio wave can make it 100 miles out and 100 miles back in about 1/1000th of a second.

In contrast to this, an acoustic wave travels at approximately 1,100 feet per second or roughly 11 nautical miles per minute, depending on the temperature. At this speed, it would take a very long time for a sound wave to hit its target and bounce back to the receiver. For this reason, sound radiation is limited to marine applications underwater, where sound actually travels four times as fast as it does in air, where light can't penetrate, and where distances are measured in feet, meters, or fathoms, not miles.

Can Radio Waves Hurt You?

Earlier in this article, we reported that Ralph Sponar

III of United Radio, Inc. said he stays out of the beam whenever he's working on a radar. But where, exactly, is the beam.

According to Buddy Morgan, Product Manager for Simrad, Inc., in Lynnwood, WA, the beam extends 25 degrees vertically, which is 12.5 degrees up angle and 12.5 degrees down angle.

Indeed, there are situations where mariners are working in the beam. For example, west coast fish boats and some east coast tuna boats, might have a flybridge with a radar mounted just forward of the helm station on the roof of the wheelhouse. This places the helmsman on the flybridge in the beam. On larger boats, with the radar mounted low on a mast, a man standing watch on the wing deck might be in the beam. Does this expose the to dangerous radiation?

"When I first got into this business," says Morgan, "I tried to find research on the subject. There wasn't any."

And there still isn't, not in terms of risk assessment. For example, we know the risks associated with various exposure times to ionizing radiation, but we don't know the risks for various times of exposure to non-ionizing radiation. How long does it take for a person to develop a skin cancer from being out in the sun? Nobody knows. We know there's a potential danger there, but that's it. We know even less about short and long term exposures to radio waves.

So, how close should you get?

"What the manufacturers usually quote in the manuals is 3' to 6'," says Ken Englert, Owner of Maritime Communications, Inc., in Marina Del Ray, CA. "Common sense tells us to stay out of harms way. But we tell our people that 6' to 8' is well beyond the manufacturer's recommendations, and I've had people tell me you have to have the dome off and your head stuck way up inside against the antenna before anything can happen, and even then it's a matter of exposure time."

Exposure time adds an interesting twist to this question, because a radar antenna is always turning, and with it, the transmission.

This is an important point because, as we mentioned before, non-ionizing radiation must be absorbed in order to cause damage. So, if you're only being hit for a fraction of a second on every rotation of the scanner, then the absorption rate is considerably less than it would be if you were exposed to a continuous stream of radiation. In other words, there's a 'cool-down period' that can be taken into consideration.

"I think I'd be more concerned about holding a cell phone to my head," says Morgan.

Another important consideration is that no matter how hard you try, you can't stand close to the more powerful radars because you'd be knocked off your feet by the turning array. If a safe distance is 3', and you're standing next to a 6' array, you're already standing a minimum of 3' away.

"The rule of thumb is that if you can't get hit with the scanner array, you're safe.

What happens if you stop the scanner from turning and walk inside the radius? "95% to 100% of the radars out there won't transmit when you stop the scanner," says Sponar Jr., "unless the scanner is stopped by a technician or a [computer] program."

Still, everyone pretty much agrees you're better off staying out of the beam. Be wary of short exposures to the older, more powerful S-band radars, which could be transmitting in the 50 to 60 kW range, and also watch out for long term exposure to the smaller radars. In fact, a lot of equipment now has the capability of shutting off transmission for certain sectors. If you're sitting on the flybridge behind a scanner, you can set the radar to top transmitting through that particular angle. You won't see any returns from that sector, but it will be behind you.

Meanwhile, to varying degrees, it pays to be wary of all the radiation on a boat, from the UV purifiers down below, to the SSB and satellite communications antennas above. The latter, which always point in the same relative direction, also transmits in the microwave portion of the spectrum.

What's the bottom line? Perhaps Jeff Chiampa, a marine safety officer with the Portland, Maine U.S.C.G. Marine Safety Office, has the answer. According to him, the U.S.C.G., which gets a lot of its guidelines from the military, makes the following recommendation:

[From the Navigation and Vessel Inspection Circular (NVIC) 586 of t

"It is hazardous to work in the vicinity of radio antenna or radar scanners due to the danger from both radiation and the rotation of scanners. Permission should be obtained from the person in charge of the bridge before such work is undertaken, and the circuit fuses should be removed to make the equipment inoperable. A man aloft notice should be placed on the equipment."

USCG ENGINE REPLACEMENT

CHAPTER THIRTEEN
RE-POWERING

Tired Motor? Now What?
For the majority of boat owners, the motivation for re-powering is a search for added horsepower. For others, it's a matter of reliability and a cleaner bilge. Others want to go from gas to diesel. Whatever the reason, the job doesn't always go as smoothly as planned.

Even if you try to replace an old engine with a new version of the exact same model, chances are it will have a different footprint, profile, and power curve. Does this mean you'll have to refit everything, e.g. engine beds, compartment dimensions, shafts, through-hulls, plumbing, stuffing boxes, stern tubes, and exhausts? The answer depends on a lot of things, not the least of which is why you want to re-power in the first place.

Do You Really Need All That Horsepower?

Dick Newbegin of Camden, Maine, is a former Chief

Engineer for an Alaskan factory trawler. He's now dragging for scallops and urchins in his spare time and working full time as an engineer for a large marina that services multi-million dollar yachts. A graduate of Maine Maritime Academy and a long time commercial fisherman, he believes that "concept" is the place to start when looking for a new engine.

"I had a 38' Young Brothers with a 210 hp Caterpillar 3208N that I used for scalloping and gillnetting. When that engine got tired, I decided to re-power with a 3208T, because I wanted to go faster. The specs said the 1-1/2" diameter stainless shaft could take the extra 90 hp and a lot more, but what it says on paper and what happens in real life are two different things."

Newbegin soon discovered the hard way that the shaft had an 11/16ths" droop over it's 5-1/2' span from the stuffing box to the coupling. He went on to learn that the 3208N had been installed without ever compensating for this droop. At 210 shp, it didn't make a difference, but at 300 shp, it made a big difference.

"There was a vibration that manifested itself as a buzzing under the heels of your feet. Eventually, the vibration crystallized the inside of the stainless shaft 1/2" just inside of the propeller hub. After breaking the second shaft, I seriously questioned the whole engineering. I ended-up installing a 1-3/4" stainless shaft and aligning the engine to compensate for the droop. I also had to rebuild the beds because the original bolts that held the angle iron to the longitudinals had worn to the point where they were 3/16" below where they should have been."

Obviously, one of the most important steps in any re-power job is the alignment of the engine. Once the foundation is rebuilt and the engine set into position, it's time to fasten the output and propeller shaft couplings. Newbegin recommends this be done twice, first, while the boat is on the hard, and then again, when it has been in the water for awhile and has had a chance to acclimate to its given operating environment.

On the jack stands, the inside stuffing box and stern tube (or box and strut) may not line up properly, and any misalignment between the inboard and outboard sections of the shaft will cause it to flex abnormally. Wood, and even fiberglass, are very flexible, which is why the best alignment will be achieved only after the boat has had a chance to settle into it's correct shape.

With the new engine bedded and aligned properly, the 38' Young Brothers ran smooth as a smelt, but Newbegin now believes that re-powering the old boat with a bigger engine—for the type of fishing he was doing—just wasn't cost-effective. Consequently, his attitude has changed slightly over the years, which leads him to question the logic behind a lot of today's re-powers. "There are limitations," he says. "Sure, in most cases, a bigger engine will allow you to go faster, but at what cost? You burn more fuel, decrease your range, and shorten engine life, because, basically, the candle that burns twice as bright burns half as long. Sometimes, just sticking to the original blueprinting of a boat is the way to go."

Newbegin has a different boat now, a 38' Novi-style dragger powered by a 135 hp Ford diesel and a 3:1 reduction gear.

Stewart Tuttle, a Marine Sales Representative at Southworth-Milton Power Systems in Scarborough, Maine, is another one to question the logic of many of today's re-powers. He says some fishermen have reached the point of little or no returns. "They put in a lot of horsepower, but it ends up being so much weight, their boats actually go slower."

Making it Fit

When it comes to finding a drop-in replacement, there's good news and bad news. The good news is that diesel engines are better than ever, providing more power from smaller packages. The bad news is that the simple drop-in re-power is virtually a thing of the past. New diesels have slightly different configurations, accessories, and power curves.

"Everybody has it in their mind that their new engine will be easy to bolt in ," says Tuttle. "But there have been lots of changes over the years. For instance, the footprint might be the same, but not the weight. Another example is the instrument panel. Our old instrument panels [for Caterpillar diesels] were mechanical, with pressure lines leading from the engine straight to the gauges."

Meanwhile, whether it's a drop-in replacement or a re-power with a bigger diesel, Jonathan Parrott, Director of Engineering at Jensen Maritime Consultants in Seattle, recommends that every fisherman conduct a torsional vibration analysis when matching a new engine to the old gear, bearings, and shafting. This means going through the entire drive train and checking everything for critical frequencies, a four to eight hour job for an engineer. "You want to make sure you don't have a vibration—particularly at cruising speed—that might cause a problem."

Obviously, vibration is to be avoided at all costs, because it can lead to mundane problems as well as catastrophic failures; Dick Newbegin's old boat is a perfect example of how vibration due to misalignment can turn around and bite you in the lazarette.

For example, take the gearbox. Say you remove a particular 700 hp engine and replace it with another one of a different make. More than likely the reduction gear from the old engine will work for the new one, but there's also the possibility it won't. "A gearbox rating is not only a horsepower rating," says Parrott. "It's a speed rating. So a specific gearbox might have a 700 horsepower rating, but it might not be at the rpm you want."

Ideally, your engine sales representative will know the makes and models of compatible gearboxes and engines. But this doesn't negate the value of torsional vibration analysis. Whether you have a matching gear or not, potential problems may exist that you can't hear or feel. It's like a big puzzle, and all the pieces have to fit together perfectly, starting with the biggest pice and going all the way down to the smallest.

"The most significant problem is space," says John Gilbert Jr. of John W. Gilbert & Associates (Naval Architects) in Boston, Mass. "When it's a straight re-power, height and width can be critical. But alignments really get tricky when we cut the boat in half. For us, re-powers are often coupled with mid-body extensions, and if you're lengthening the boat, you have to change the shaft angle. Sometimes there isn't enough headroom for the engine to be realigned. Sometimes we have to put in a new foundation. If it's a smaller boat, it may not have the depth. In some cases, finding space for the silencer can be very difficult, because of the engine's minimum [allowable] exhaust back pressure. An engine's silencer should be in the first 2/3rds of the run."

Space, engine alignment, exhaust, reduction gear, shafting, bearings, compatibility...what else is there to consider?

"The other things we usually look into," says Gilbert," are the fuel lines, pipe sizing for cooling

This last point is probably one of the most significant for fisherman who want to re-power with a bigger engine. Because while you can add larger diameter fuel lines, through-hulls, exhausts and cooling pipes fairly easily, you can't swing a bigger prop if the boat won't accommodate it. And this brings us back to Tuttle's and Newbegin's earlier comments. You have to consider the boat's blueprinting, and ask yourself the question: Why am I doing this?

CAN YOU SAVE THE ENGINE IN THIS BOAT?

CHAPTER FOURTEEN
SAVING A DROWNED ENGINE

What to do?
When the Lorraine II collided with a concrete barge in Maine's Penobscot Bay, there wasn't much time for dillydallying. A 4' hole in the stem was letting the sea in by the bucketful. Fortunately, the captain had the presence of mind to ground the 50' fiberglass dragger; he pinned the throttle and made for the nearest beach. It was a damn good plan, and it would have worked, too, had it not been for a few boulders planted during the pleistocene period. The rocks ripped out sections of the keel directly under the engine room and fish hold, and they tore through an area of the hull at the turn of the bilge. The boat made it ashore, but her bottom had been seriously compromised. When the tide came and went, she was almost as good as sunk.

Henry Lindahl, a lobsterman from Owls Head, Maine and the owner of the dragger, supervised the salvage. He hired Bernard Shaw and his crew at Shaw's Yacht

Service, and on the sixth day after the sinking, on an early morning high tide, the boat was lifted half out of the water (with emergency pumps at full tilt and officials of the U.S.C.G. standing-by), tied to the bow of the barge, and pushed three or so miles to the nearest slipway. As is most often the case with these types of accidents, a marine surveyor was immediately called to assess the damage to the boat and its machinery.

The bottom line? The insurance company deemed the Lorraine II a total loss and Lindahl got a check for the coverage amount. But that wasn't the end of the story. Because of all the time and money he'd spent on the salvage operation, he was given additional compensation for the work he had done personally.

Interestingly enough, Lindahl ended-up with the boat. The company offered it as compensation and he thought it was a fair deal. Rough as the vessel was, he believed it could be repaired privately as a "project" boat, i.e. at off-hours rates and not shipyard rates. In addition, Lindahl knew the vessel had a lot of good equipment on it...a new prop, new net reel—and Cummins 855 diesel engine with low engine hours.

In fact, it was the Cummins, or more accurately, the state of the Cummins, that helped Lindahl make up his mind. "If the engine had gone down running," he says, "I probably wouldn't have taken the boat."

Did it Go Down Running?

Art Stanley, of Art's Marine Service in Owls Head,

ME, has rebuilt a number of drowned engines. He says it's one thing for them to go down while they're shut off, quite another for them to sink when they're running. If an engine goes down running, and it has a chance to suck water into the manifold, and worse, the cylinders, costs to repair the engine skyrocket.

Adam's Rib was a steal dragger with a 3408 Caterpillar that sank while running at about 1,000 rpm. Sucking in seawater at that speed created a compression shock that not only bent the rods but also snapped the engine's head bolts. It was a situation that obviously left Stanley and the vessel's owner with too many questions about the engine. Was the crank bent? Were there hairline fractures in the heads? Just too many doubts. "We ended-up putting in a new 3406," says Stanley.

Fortunately, this wasn't what happened at the time of the Lorraine II accident. According to the captain, Mike Williams, he shut the engine off before the worst of the flooding occurred—good news for the Lorraine II and Lindahl.

"Everybody told me I had to take care of this right away," he says. "But the sinking came at the worst time. It was the beginning of the [lobster] season and I had about 300 traps out. And everyone was getting their boats ready. I had a hard time finding a diesel mechanic."

Lindahl didn't have the experience as a mechanic to do the whole job himself, but he did take the time to pickle the engine. "You have to pickle the engine," says

Dave Kennedy, Sales Manager for New England Detroit Diesel/Allison in Portland, Maine. "Salt will leave deposits on internal castings. There's also going to be salt deposits on rings, valve stems, cylinder liners. You have to get that out."

Bruce Murgita, a mechanic for the O'Hara Corporation in Rockland, Maine, explains the pickling process as follows: "What we like to do," he says, "is drain the oil and flush the lubricating system with fresh water, then fill it with diesel and turn the engine over by hand." This gets all the moving parts of the engine saturated and helps displace the water.

In addition to filling the oil circuit with diesel fuel, mechanics will also remove the injectors and fill the cylinders—but they'll stop short of filling the fresh water cooling circuit with fuel. "As long as it doesn't have a cracked hose," says Murgita, "it's a closed system and shouldn't have any salt water in it."

So Lindahl drained the oil and coolant. He removed the injectors, flushed the cooling circuit, and filled the appropriate voids with diesel fuel. He should have started work on the engine right away, but without a ready, willing, and able mechanic in the wings, his hands were tied.

"The most important thing to do is take care of the engine immediately," says Stanley. "Get it running and dried out."

Stanley believes the best way to get the water out of

an engine is to 'bake' it out. He pulls the injectors and rolls the engine by hand. "not with the starter," he cautions, "because the point here is to get the water out before cranking the engine." He replaces all electrical components and all wiring, and breaks down the exhaust system. Best possible scenario is to refill the engine with oil and coolant as soon as you can—using what was called for in the first place—and restart.

"Ideally," he says, "you want to put the engine back in service in a boat as soon as possible. It's not the same to have it on the blocks in the yard running."

Stanley adds he might run the engine for ten minutes and change the oil, or he might run it and change the oil three more times. It all depends on whether or not he sees any signs of water or contamination. "You really have to cook the water out of it," he says. "Most engines have a crank vent system, and this will help get rid of it."

A few other things to consider, according to Stanley, are where the boat sank, its orientation under water, and for how long it was underwater.

Meanwhile, without the dirt, mud, and electrolysis of a thoroughly drowned engine complicating matters, trying to restart an engine that didn't sink while running is not all that daunting a task. At least it wasn't for Lindahl, who finally found a mechanic to take care of the Lorraine II's Cummins. "Neal Pottle, a diesel truck mechanic, came down from Palermo," says Lindahl, "and after about 12 man hours of work, we had the engine up and running. When it started, it started beautifully."

Electronic Diesels
Electronic Diesels Aren't That Much Different.

Nobody buys an engine with the notion that one day they'll have to bring it up from the deep and restart it, but there is some concern out there that new electronic controls require much more in terms of major repair. However, according to Dave Kennedy, Sales Manager for New England Detroit Diesel/Allison, in Portland, Maine, this concern is really unwarranted, because the basic components of the engines are still the same.

In the event of a sinking. Treat the engine exactly the same as if it were a fully mechanical diesel. Pickle it as described above. Electronic components are another story. Modules can be tested, appraised and swapped out, if need be. Sound expensive? It didn't used to be, but these things are getting quite expensive. In most cases, the costs for testing and repair of the modules and solenoids are not at the level where they would break the bank. On the other hand, display heads for electronic controlled diesels can run in the thousands, and so can unit injectors. However, although no one can say for sure what will happen to components in the even of a sinking, engine manufacturers are quick to point out the ruggedized and waterproofed features in today's modern, electronic diesels.

ZEISS CUTAWAY

CHAPTER FIFTEEN
SEEING IS BELIEVING

Binoculars

There's no question that seafaring is hard on the human body. Over a period of time, things start to go, not the least of which are the eyes. Sun, glare, salt, wind, all contribute to poor vision. What can you do? Wear wraparound sunglasses and a ball cap for one. For another, carry a good pair of binoculars.

Porro Versus Roof

Binoculars are basically two telescope bodies mounted side by side. But they're also a bit more. In addition to the magnifying lenses in telescopes—used to enlarge and focus the image—binoculars have a prism-erecting system. Without prism systems, images would appear upside down and backwards, and binoculars would be much larger.

For the most part, it's the prism system that separates

the two main classes of binoculars. The Porro prism—designed in the 1800's by an Italian inventor named, Porro—is very light-efficient. It provides excellent image contrast and continues to be a major player in the industry.

The other system, the roof prism system, gives binoculars an H-shape—a straight body in which the outer set of magnifying lenses (the objective lenses) are directly in line with the the inner set of lenses (the eyepieces). Roofs allow for lighter, more compact binoculars, which, over the years, have made them the focus of a lot of R&D. In fact, even though roofs are characteristically not as light-efficient or as good at providing image contrast as Porros, they've evolved to a point of dominance in the industry.

"Porros were the most popular until phase-shift coatings [P-coatings] were invented," says Karen Lutto, a Public Relations Representative for Zeiss Sports Optics in Petersburg, VA. "With P-coatings [invented by Zeiss in the 30s], roof prisms have become optically comparable to Porros."

They also cost more, and this sometimes means that cheap roof prism binoculars do not have the definition, contrast, and low light capabilities of similarly priced Porro prism binoculars. "Without the coatings," says Lutto, "a roof prism is just not as good as a Porro."

Meanwhile, not all Porro prisms are created equal. The better prisms have high-density glass (BAK-4) and the less expensive prisms have lower density glass

(BAK-7). You can tell which one is which by holding the binoculars away from your face and looking in the eyepiece. The better prisms have rounded exit pupils, and the lower quality ones have exit pupils with squared edges...a good thing to remember if you're thinking about buying a used binocular.

Leica and Swarovski use roofs exclusively. Bausch & Lomb, Nikon, and Canon, use roofs systems in the their top of the line models (roughly $800 and up) and Porros in their lower priced models. And Zeiss, kind of the exception, uses roofs in all of its binoculars, except the 20 x 60 S (suggested retail about $6,000), which uses Porros.

Brightness and Fuzziness

Ever look through a window and see your reflection? Of course you have. This happens because glass surfaces reflect some of the light that hits them. In fact, as much as 5% of the incident light bounces off even the best untreated optical glass surface.

In a binocular, there might be 14 or more surfaces, half of which are facing the inside. As light reflects, it starts bouncing around the tubular housing. This is particularly true for older binoculars and low quality new ones. Think of it as the inside of a movie theater. The lighter the theater is, the harder it is to see the movie.

Obviously, if you want to see the movie better, you can brighten the images on the screen. Binocular manufacturers do something similar by increasing the size of the objective lenses. This compensates for the reflectivity of the glass by allowing more light in. Unfortunately, increasing the size of the lenses adds significant weight and bulk to the binoculars. making them virtually impossible to use without a tripod.

Today, with quality optics and the evolution of special multi-layer coatings, high-end binoculars have transmission efficiencies of 95%. But don't take this for granted. Shop around and ask plenty of questions, because there are coatings, full coatings, multi-coatings, and full multi-coatings, the latter meaning that every air-to-glass surface is multi-coated.

One way to check and compare the optics of a few different instruments is to set up your own little bench test. Focus on a back wall. What you want is no the least contrast and distortion from dead center all the way to the outer edges of your field of vision.

The Numbers Game

7 x 50. What do these numbers mean, anyway? Simple.

The first number is the number of times the image is increased, otherwise known as the power or magnification. In other words, at a magnification of 7x, and object will appear 7-times closer than it really is. If it's 200 yards away, it will appear as if it's 28.6 yards away. The second number is the diameter of the objective lens in millimeters, which determines how much lights enters the binoculars. A third value, the diameter of the exit pupils, can be extrapolated from the first two. Just divide the diameter of the objective lens by the power of magnification.

This exit pupil diameter is kind of important, because the larger it is, the better the binoculars will perform under low-light conditions. They're like the pupils in your eyes. The wider the diameter, the more light they allow through. Unfortunately, there is a caveat.

Earlier we mentioned that aging causes the human machine, particularly the eyes, to wear down. In fact, one of the first things to go is night vision. When you're young, your pupils have the ability to open to 7 mm, but as you get older, you lose that ability. A teenager might have great night vision, but when he reaches 50 or 60 years old, he might only be able to open his pupils to 5 mm. This means that a binocular with a 7 mm exit pupil is essentially lost on an older person, at least in terms of low-light viewing. You can't fit a 7 mm peg into a 5 mm hole.

Another number worth considering relates to field of view. This is either designated by degrees or the width of the image in feet at 1000 yards. To convert between the two, just multiply the degrees by 52.5 or divide by the width of the image by the same number. In other words, the Steiner Commander III has a field of view of 385' in 1000 yards. Divide 385 by 52.5 and you get 7.3 degrees. This is a good coefficient to keep handy if you're out shopping for binoculars, because not all companies use the same measurement.

How important is the field of view? Not much if all you want to do is take your binoculars to the gun range and keep it trained on the target. But if you're bird watching, or whale watching, or trying to catch a glimpse of fish jumping, or you're searching for a man overboard over a wide area from 1,000 yards away, you'll have an easier time of it with the higher field of view.

Last but not least is eye relief. This is the maximum distance your eyeball can be from the eyepiece and still see the complete field of view. Generally speaking, the wider the field of view, the shorter the eye relief. For non-glasses wearers, it's not that significant a characteristic, but people who wear glasses should take note for obvious reasons, unless they do what I do, i.e. slide their glasses up and out of the way whenever they use binoculars.

Special Features

Over the years, manufacturers have added all kinds of special features to their instruments, including electronic bearing compasses, range finding reticles, laser range finders, gas-filled housings (to prevent fogging), waterproof components; there are even pressure tested to several atmospheres.

Bearing compasses and range-finding reticles (trigonometric scales) are probably what most people think of first when they look for extras. The compass is usually designed so you can see it through-the-lens, while the reticle shows up as cross hairs in the eyepiece.

Using a reticle is a simple matter of multiplying and

dividing. Take the known height of the object, say a 60' tall light tower, divide it by its height in reticle units, say 12, then multiply by 1000. The resulting distance (in this case 5000') is in whatever units the actual height of the object is. It could be feet, meters or yards, it doesn't matter.

Leica's Geovid goes one giant step further. The Geovid has an infrared laser

But stabilizing systems, for my money, are the most sophisticated and interesting inventions of all. They're high-tech optical add-ons that maintain focus despite varying degrees of hand-tremor.

According to Bruce Cavey, Vice President and General Manager of Zeiss Sports Optics, the Zeiss 20 X 60 S uses a piece of steel cut into layers and formed into a kind of Slinky. The Slinky works in conjunction with a prism carrier. Cavey says the 20 X 60 S will resolve the image of a dime at 100 yards. "You wouldn't be able to read the writing on it," he says, "but you'll know it's a dime."

Another stabilized binocular is the Canon 15 X 45 IS. It uses flexible, oil filled bellows between two transparent plates. As the compass tilts and pitches, the instrument—called a vari-angle prism—works to keep the image in focus.

What does the future hold in store for binoculars? Industry leaders won't say. But in the years to come, I'd be looking for a marriage between high quality optics and all kinds of existing technologies, e.g. video, audio, laser, photoelectric amplification, even remote positioning. Who knows? Not too long from now, you might be able to analyze images at a distance and note their positions in lat/long.

###

Marine vs Field

According to Terry Moore, Director of Sales and Marketing for Leica Sports Optics in Northvale, NJ, there are no required standards for the use of the words, marine, weatherproofing, or waterproofing. What means one thing to one company could mean a completely different thing to another company.

"For example, all of our full size binoculars are waterproof to a depth of 16-1/2' (1/2 atmosphere)," he says. "But waterproofing does not mean emersion to all companies."

Moore goes on to explain that waterproofing is not just a matter of preventing moisture from getting into the instrument from the outside. It also means preventing condensation from forming on the inside. To this end, manufacturers use different methods to keep the inside of the instrument as dry as possible.

Is there a difference between the degree of waterproofness between individually focused lenses and center-focused lenses, as some fishermen believe? Not any more, because innovations over the last fifteen years have made center-focus instrumentation as waterproof as individual-focus. "We brought waterproofing to center-focus with the Ultratrinovid system," says Moore, speaking about Leica's models. "Other companies, Pentax, Canon, etc., have done something similar."

One more characteristic that often applies to marine binoculars is the diameter of the exit pupil. Because a larger exit pupil improves low-light viewing and image stability, it's very common for many marine binoculars to be in the 7 X 50 category. A 7 X 50 binocular allows for about a 7 millimeter exit pupil, which is about the largest you can get.

Is there a bottom line in all this? Sure. Don't buy binoculars based solely on the words, marine, waterproof, and weatherproof. Instead, base your decision on performance, individual features, the all-important warranty, and your personal taste.

###

Night Vision Scopes

What, exactly, is a night vision scope, and does it have any practical value to the mariner? The first question is fairly easy to answer. The second is not.

Regarding question one; Night vision scopes or light amplifiers are devices that work because of the photoelectric effect, a principle of physics that describes how electrons can be freed from matter by certain wavelengths (frequencies) of electromagnetic radiation. It sounds wordy and complicated, but it's really not that difficult to understand.

A light amplifier receives a low light image through a

series of optics set in a tubular housing. The light energy from this image is converted to electrical energy via a photocathode; Basically, individual photons (particles of light) liberate a proportional number of electrons after interacting with the atoms on a thin metallic plate.

These free electrons are channeled and amplified through the image intensifier, a very sophisticated device that has gone through three major evolutionary changes over the last forty years. But more about that later.

After being amplified, the electrons are projected onto the phosphors of an anode. Electrical energy is converted back into light energy and the image is reconstructed, a little like it is on a television.

Night vision systems got a lot of media attention just before, and immediately following, the U.S. military incursion into Panama in 1989. Senators and congressmen began scrutinizing the high-tech scopes when it became known that a supposedly inordinate number of helicopter accidents had occurred as a result of pilot error; It was claimed that aircraft flying in formation at close quarters were colliding because the helmet fitted night vision devices seriously reduced a pilot's peripheral vision.

This particular media blitz gave most people their first look at the view through a light amplifier. Photographs in newspapers and film footage on television specials and network news programs showed low light images as seen through available military hardware. If you remember, the images, though colorless, were extremely well defined and clear.

Of course, night vision systems are like most other pieces of advanced, high-technology equipment; The military gets the absolute best, and we buy the rest. It's not that the Pentagon makes the technology completely unavailable (actually they do in a lot of cases) but usually, when industry breaks new ground with a particular device, its predecessor becomes more available and more affordable to the public. With regard to night vision systems, industry is stepping up to the eight rung on the ladder, which means that most of us will be buying, at the top of the line, what are considered Gen III+, or as developers and procurement specialists like to say - Gen-3 Plus.

Now, if you're not about to pay $5,000 or more for a Gen-3-plus scope, you'll probably end up buying Gen-1-plus-one, Gen-2, Gen-2-plus-one, and Gen-3 equipment.
Here's a quick breakdown:

Gen-1 light amplifiers - the original Starlight Scopes - were designed sometime during the 1950s. They typically have multi-stage intensifier tubes with low resolution capabilities and a propensity toward glare burn-out. Though they can still be purchased, a buyer should be wary. Many of the bargain scopes currently being marketed have nothing more than rebuilt tubes that have been tweaked beyond their operational limits. These devices, while comparatively inexpensive, will not last long.

Other Gen-1 devices, like Fujinon's recently

introduced Starscope, have Gen-1-plus-one tubes that, according to Fujinon, rival the capabilities of Gen-2 based equipment.

Gen-2 technology introduced what is referred to as the microchannel plate (MCP). This eliminated the need for multiple stages and, at the same time, reduced the streaking, distortion, washout, blooming, and vignetting that was typical of most first generation tubes. Basically, the MCP improves the overall contrast of the image and makes it possible to look into a lighted subject without losing the image or burning out the tube; (i.e., the Fujinon Starscope shuts itself off when too much light enters the tube.)

Gen-2-plus-one and the more advanced Gen-3, which uses a gallium-arsenide image intensifier, as well as Gen-3-plus-one, are available from a relatively small family of night vision system manufacturers. These devices, though expensive, provide civilians with the very best light amplification on the market.

What Good Are They?

The purpose of a night scope, as originally conceived, was to provide the stealthy observation of a target. For this reason, they're particularly useful to law enforcement, search and rescue, and military personnel. In fact, most light amplifiers are designed to be fitted to weapons systems, helmets or cameras.

Despite this however, night vision systems are being actively marketed to the marine end-user, yachtsmen, work boat operators, and commercial fishermen alike.

Obviously, someone has foreseen that such equipment can be useful aboard a vessel

There's no doubt that a device with the power to see through the dark, or more accurately, see reflections in star light, is a valuable tool. But there are alternatives for a good size vessel. Obviously, for most boats, both pleasure and commercial, a powerful search light and/or a radar takes precedence over a night vision scope.

On other hand, while a search beam is all right for picking out navigational markers and the end of jetties and breakwaters, it's not well suited for picking out other boats. In fact, it is against the law to direct the beam of a light into the wheelhouse of another vessel when its underway. Bearing this in mind, a night vision scope, in high traffic areas, can be a good supplement to a radar.

There is one other very important point to consider. Because night vision systems are considered to be advanced pieces of military equipment, their distribution is regulated by the State Department's Office of Defense Trades Control. This means that any optical device on the U.S. Defense munitions list cannot be legally taken out of U.S. territorial waters without an export license. Currently (01/26/14), all second generation and above night vision systems are on the munitions list.

LANGE DOUBLE RAM STEERING SYSTEM

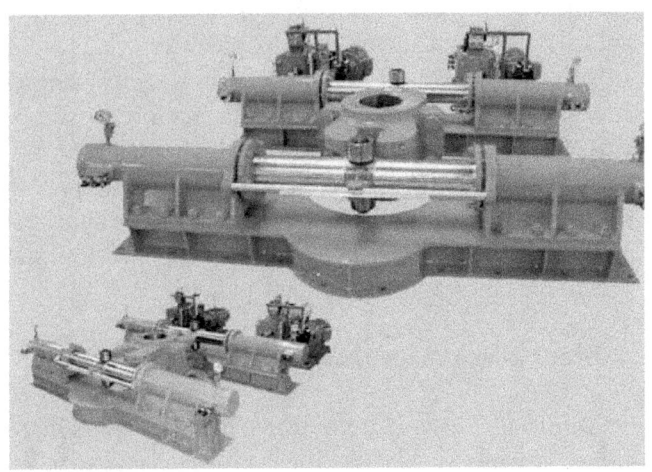

CHAPTER SIXTEEN
STEERING

Steering Systems

Whatever type of boat you have, longliner, lobster smack, factory trawler, tramp steamer, yacht, etc., it needs some form of steering system. The simplest is mechanical, whereby a rack and pinion drive, worm drive, or chain and sprocket and/or pulley and cable transfer motion from the wheel to the tiller arm and rudder. The next level up is the manual hydraulic system. After this, there's the hydraulic power steering system, which uses a motor-driven pump.

Mechanical Steering

You don't see too many fully mechanical systems on boats anymore, unless they're on outboards, small inboard boats, older vessels like west coast salmon trollers or halibut schooners, and older yachts—or they're installed as emergency steering systems, a tiller modified to fit over the rudder stock, or a block and tackle system for emergency steering on an old tug or sailing vessel.

The downsides to mechanical steering systems are numerous: First of all, the force of the water pressure against the rudder is transferred directly to the drive system and, consequently, the helmsman's arms, which, in rough seas, can be a lot of force. It's also a lot of force with twin outboard or outdrive installations, regardless of sea conditions, because with these types of drives you're trying to move the entire propulsion unit.

In addition, mechanical systems can't be too long, or be shaped to go around too many turns. Older systems can't keep the vessel on a given heading if you let go of the wheel, because under most conditions, if you let go of the wheel without locking it in place, the loads on the rudder(s) will eventually overwhelm the system. This includes the propulsion side loads (prop walk effects) of your propeller(s). Today, however, many mechanical steering systems are offered with a no-feedback option, basically a mechanical clutch that compensates for the forces on the system. Most of these NFB systems, as they're called, are designed for outboard powered boats under 35' in length.

Hydraulic Steering

In manual hydraulic steering systems, a helm pump is connected to the steering cylinder via hose or pipe. As you turn the wheel one way or the other, the pump pressurizes the fluid and causes the piston in the steering cylinder to move in and out, which, in turn, moves the tiller to port or starboard. With this system, feedback is eliminated by a pair of check valves built into the helm pump.

In a power hydraulic system, the engine or motor driven pump keeps the circuit pressurized and the helm is just a valve that controls the flow. When you turn the steering wheel or jog stick, you open the flow to the steering cylinder. In this system, feedback is controlled by the valve.

Manual hydraulic steering systems are limited to boats under about 50' in size. Bigger than that, and you start to getting into some problems with the number of turns it takes to go hard over to hard over. In other words, because the operator is the prime mover, the fewer the turns you want from hard over to hard over and/or the bigger the boat, the harder it will be to turn the steering wheel.

The thing to remember is that the cylinder does the work. When you turn the wheel you create flow, and that flow overcomes the resistance of the load. This means is that if you have a boat over 50' and you want less than 5 or 6 turns of the steering wheel to go from hard over to hard over, the only way to get the necessary pressures is with a power hydraulic steering system. Otherwise, there's just too much work involved turning the steering wheel.

Sizing and Matching a System

According to John Richardson, Sales Manager for Marine Hydraulic Engineering Co., Inc., of Rockland,

Maine, manufacturers of Hydro-Slave Hydraulic Power Steering Products and the world's largest seller of hydraulic power steering, the system has to be matched for the application. "Pumps, cylinders, and steering helms," he says, "are all matched so the system operates correctly. We see problems when people mismatch components and operate with an improper pulley ratio on their power steering. Many people are turning their pumps too fast."

Basically, you start from the rudder and work forward. The steering cylinder is sized to overcome the force against the rudder. That's the important part. You pick a helm pump [for manual systems] or a steering helm [for power systems] that will give you the number of turns you want to have based on what size cylinder you need. If you were lobstering you'd want fewer turns and a quicker response.

Marine Hydraulic Engineering Co., Inc., of Rockland manufactures belt-drive pumps as well as pumps that mount to an auxiliary drive on an engine. They have five classes of systems that use three different size steering pumps, depending on the length of the boat. "We control the number of turns by the cubic inch displacement of the steering helm," says Richardson. "In other words, the larger the displacement, the quicker the boat will turn. When they [customers] call up, they ask for a faster or slower turning helm. We can go anywhere from 1.8 turns on the small class to as high as 6 turns on the large class. The logic behind this is that you don't want a 120' boat turning on a dime."

Another consideration is the rudder itself. "We use a standard formula based on size and shape of rudder and speed of boat," says a steering system professional from Europe. "Also, when we're choosing a steering system, it's important for us to have an accurate measurement of the counterbalance of the rudder. Counterbalance is the material forward of the centerline of the rudder post. It improves steering by using more of the prop wash and hull wake."

Meanwhile, these hydraulic systems are designed specifically for marine use. They're made of bronze and stainless steel and have waterproof and saltwater resistant seals. They're also balanced. For example, a regular cylinder has one rod coming in and out the barrel of the cylinder. A balanced cylinder is one where there's a dummy rod coming out the other end. The dummy rod makes it so the volume is the same whether that cylinder is extending or retracting. On a regular cylinder, it takes more oil to extend the rod than it does to retract it, because part of the volume is already used by the rod. An unbalanced cylinder would require a greater number of turns in one direction than the other."

Bear in mind, not all dummy rods are visible, because some manufacturers have them enclosed in a housing that's part of the steering cylinder. Other manufacturers use two regular cylinders in a push-pull arrangement, which not only accomplishes the same thing—balances the system—but also adds a level of redundancy and therefore safety. A dual ram steering system will provide a greater and smoother force on the tiller arm.

Autopilots and Other Add Ons

Whether you use a manual hydraulic system or a power hydraulic system, you may someday want an autopilot. Power hydraulic systems are ready made for this: They're already motorized, i.e. there's either an engine-driven pump or an electric motor driven pump. These systems can support a variety of electronic autopilot interfaces. Simply add a unit with an autopilot solenoid valve, built in flow control, and double check valves.

This, of course, is an added benefit of the power steering system...it's easily adapted to electronic controls. In fact, on large boats, the steering helm is completely electronic. Instead of a steering helm with flow control valves and hose or pipe leading from the wheelhouse to the lazarette, you just wire the helm to a pair or set of solenoids. The solenoids—which can be anywhere but are usually in the engine room—push and pull on the flow control valve for the cylinder and the tiller arm.

By contrast, manual hydraulic systems need everything the power systems need—e.g., CPU, display unit, fluxgate compass—plus they need their own pump. You could also install a mechanical system, including a linear actuator attached to an auxiliary tiller arm, but this would operate independent of the hydraulic system.

Finally, keep an eye, and nose, on the hydraulic oil. If it smells bad, or looks discolored, or the filters keep getting clogged, chances are something is wrong in the system, and that 'something' could cause you to lose steering at a most inopportune time. Rigging an emergency tiller to fit over the rudder stock of a boat is not that big a deal, and it may save you from going aground or having to get towed. In an emergency, you just pull the pins connecting the hydraulic cylinder rod to the yoke on the rudder, mount the emergency tiller and you're back in business.

RULE 4000

CHAPTER SEVENTEEN
TENDING THE BILGE

Pumps

When it comes to the relationship between their boats and the sea, fishermen everywhere think the exact same way: They would very much prefer that the water on the outside of their vessels stay just where it is. Unfortunately, this is virtually impossible. Because even though the seams may be tight or the welds look like melted butter, even though the packing glands are dripless and the through-hulls, hatches and sea chests seal to perfection, a certain amount of seepage is inevitable. Our answer to this: The centrifugal electric bilge pump, a marriage of technologies that has saved thousands of vessels.

But, believe it or not, it's possible for a bilge pump and hose to sink a boat. More than a few vessels have met their doom because of improper or haphazard installations. The combination of a loose outlet hose dangling in the ocean and/or a bad float switch, a rotted or stuck check valve, no check valve at all, or no anti-siphon loop, and instead of seawater being pumped out of the bilge, it's being siphoned into it. Oops!

Which is why it's so important to pick the right pump and wire and install it correctly.

Choosing a Pump

Choosing a pump is a decision left almost exclusively to the boat builders. but there is a general guideline that most builders use. They need to have a 100 gallon per hour (gph) capacity per foot of boat at open discharge. Open discharge is the rating of the pump's output without the hose attached, i.e. with the water shooting right out of the pump. This is a perfectly good rating for determining which size pump to buy. However, if you install the pump in the bilge and place the exhaust outlet high up on the hull, you won't get the same results. In other words, you have to bear in mind the output pressure (head). The higher the head, the more work the pump has to do, and the lower the output capacity will be.

The vast majority of electric bilge pumps are of the centrifugal type, meaning that they operate by virtue of centrifugal force, a force that is directed outward along a radius. Picture a marble placed at the center of a record on a rotating turntable. The marble will roll outwards until it falls off the record. In a centrifugal pump, water at the center of a spinning impeller is directed by curved blades spirally outwards towards an exhaust port.

As it happens, there are more centrifugal pumps in the world than any other.

Also, because of their size and the fact that they have to be submerged, they can't always fit into a tight bilge compartment. For example, in my old boat, the space between the keel and the hull, forward of the collision bulkhead, didn't allow for anything but the very end of a hose. Hence, for this application, as well as those requiring a lot of lift and/or head, a self-priming diaphragm or flexible impeller is the better choice.

Installation

Because the majority of centrifugal pumps have literally no lift capability, the impeller must be submerged at all times. If the pump is out of the water and/or gets air-bound (because of loops or dips in the hose), it will not work. Moreover, with any centrifugal pump, you need to run your discharge hose continually upwards in order to prevent air-lock. You may also find a situation—a low outlet near the waterline—where you might want to install a through-hull with a check valve. The valve must be installed at the through-hull, because any significant amount of water that gets trapped in the hose will not be cleared by the pump. The pump won't have the power.

Speaking of which, bear in mind output capacities of centrifugal pumps can be affected by other factors, including, ice, debris, monofilament, and improper installation. For example, if you reduce the diameter of the output hose, you reduce the output capacity of the pump. And there can be a huge difference between using smooth hose and corrugated hose. We're talking in the neighborhood of 15%. Which is why most pump engineers will call for smooth, heavy duty PVC sanitation hose with the rigid PVC helix. It provides good flexibility as well as collapse resistance and tight-bend capabilities.

Here are a few other points to remember about installing electric bilge pumps: Keep connections out of the bilge water and use liquid tape to seal everything. Secure the pump and the hose so it doesn't move around. Double-up on hose clamps. Keep the bilge free of debris. Wire an electronic counter into the circuit if you want to know how many times your pump turns on and off. And don't worry if the output is low immediately after you install a brand new pump. It takes about 10 hours for the brushes to wear in.

Maintenance and Repair

All the electric centrifugal-type bilge pump manufacturers use direct drive designs, except for Lovett Marine, which uses an indirect drive. While both should be kept clean and debris free, the direct drive pump motor—where the impeller is directly attached to the motor shaft—is more susceptible to water damage, because the shaft and motor can be exposed to water.

What this mean is that monofilament, sand, and shell hatch, can indirectly damage a direct drive motor. If the seal goes, water gets in and ruins it. Therefore, it's important to check the filter and impeller on a routine basis, because you don't want monofilament winding around the shaft and wearing against the lip seal. On the other hand, you don't want debris fouling Lovett Marine's indirect pumps either. When the motor jams or slows as a result of debris, the belts wear prematurely and have to be replaced.

The Mayfair/Johnson 2200 gph pump uses a mechanical seal instead of a lip seal. According to the company, the change was made in order to increase longevity and improve wear resistance. They wanted the seals to last as long as the motor and claim the design provides a better sealing mechanism because the seals rub against each other and not the shaft.

However, the most common problems are not the result of scarring of the lip seal or wearing of the shaft. The biggest problem is where people make their wire splices. The majority of pumps fail because of water wicking back to the pump through the wires. When the motor runs, it heats the air inside the housing. When the motor shuts off, the air cools and contracts. When it does this, you have a wicking effect.

What do you do when the motor's shot? In some cases, you throw out the whole pump and buy another, because it's more cost effective to do so. In other cases, you replace the motor. Lovett pumps can be completely rebuilt, and some Johnson and Mayfair pump motors have a unique snap-in/snap-out feature.

With a proper installation, a clean bilge, and a minimum of maintenance, an electric bilge pump can last a very long time. The two Rule 2000 pumps in my boat have been there for about eight years, and, according to Lovett,, they have pumps from the 1950's that are still in operation.

Alarms

I know more than a few boats that have been saved by high water alarms. One was headed out at full throttle when the alarm went off; turned out the gland nut of the stuffing box had backed off. The other was headed out in rough seas when the washdown hose (also the outlet for the engine's salt water cooling pump) bounced back into the boat and jammed into the engine room vent.

The first thing to remember is that while simple devices like alarms and indicator lights are pretty foolproof, the redundancy is well worth the added expense. Use both i.e., an indicator light at the helm and an alarm. And choose a float switch that you have complete confidence in, otherwise the whole installation is pointless.

There are several different kinds of float switches on the market, from simple mechanical on/off switches (where two wires make contact) to mercury switches (where a dollop of mercury makes contact), to the more advanced reed switches. There are also vacuum switches, switches with electrodes, switches with magnets, and switches that work by air pressure. Whatever type of switch that you currently use for your bilge pumps will work fine as a high water alarm switch.

Most of the switches with articulated arms can be

mounted at an angle. Others might have to be installed on their own little shelf; they need to be level to work. Obviously, you don't want to mount the switch so low that sloshing water in the bilge sets it off. Neither do you don't want it so high that you don't have time to correct the problem.

Some vessels might need more than one high water alarm. I know of one boat where the trawl cable sawed through the fiberglass transom at the turn of the bilge. The lazarette filled with water and before anyone could correct the problem, the crew was in the lifeboat.

Some companies sell fully integrated switches and alarms that are perfectly suitable for when you're on the boat, but won't cut the mustard when the boat's on a mooring. What happens if the boat's 100 yards from shore and you have a leak? There's no way that a buzzer (like your low oil pressure buzzer) or bell-type alarm will be loud enough. Therefore, you might want a more substantial noisemaker, something along the lines of an anti-theft alarm-horn or siren. They might scare the daylights out of you when you're on the boat and underway, but they'll be heard for a mile when the boat's on a mooring. Fireboy makes an alarm-horn that it sells as a separate accessory. But any auto parts store should be able to order one. Prices range from $50 to $100.

A light can help as well, and today, if you have Internet access at the marina or in the harbor, you can rig a full automated Internet based monitoring, control and alarm system.

NOVI CLASS RACE WINNER

CHAPTER EIGHTEEN
THE PERFORMANCE ENGINE

Fastest Lobster Boat
The following is reprinted in its original form. It's about the making of race engines for the 1998 Maine Lobster Boat Races.

It's a controlled calamity. A conflagration in a can. You have to be a little nuts to do it...or just plain out of your mind. But the thrill is there, borne from the not knowing as much as the knowing. Will you be a winner, or will you be bringing your engine home in a box?

Racing is a fine line between jubilation and disaster. You never know whether you're going to be in the winner's circle or the pits. Come race day, you're afraid something's going to go wrong.

During the 1998 Maine Lobster Boat Races, very little went wrong for Glenn Crawford of the C&C Machine Shop in Ellsworth, Maine. At the Searsport Race. His 502 cubic inch GM block with twin Detroit Diesel

Turbochargers took Sid Eaton's *Little Jan*, an AJ Enterprises, 26 footer, to the heart-pumping speed of 52.9 mph. That might not sound like much to a NASCAR driver or an offshore powerboat racer, but for a single-screw working lobster boat it's practically a streak of lightening.

"It's all in the combustion chamber," says Crawford. "That's where everything happens. You could sleeve the block, stroke the crank, change connecting rod lengths.... Just increasing the bore is a big help. In our case, we're increasing the volume of the chamber with forced air. Other people are doing it differently."

One of the "other people" is Tom Hofstetter at Chief Engines in Ft. Lauderdale, FL. Hofstetter and his company are famous for building high performance engines for offshore racing boats. If you're in the market to go fast you can call him up right now and choose from well over a dozen models, including one based on a 705 cubic inch Merlin block that'll deliver more than 1,400 hp. Just be ready to spend in the neighborhood of $60,000, and don't plan on feeding the thing anything less than 116 octane gasoline.

Hofstetter, who is currently building an engine for Glen Holland's 32' *Red Baron*.

Crawford, on the other hand—and many other Maine-based enthusiasts—represent the alternative approach. As owner/operators of smaller machine shops and garages, they're primary job is to keep commercial fishermen fishing. Building a lobster boat racing engine is just a hobby. But it's one that ties into a whole way of life; if these guys aren't fishing themselves, someone in their family is, or they're driving around in their father's old boat. True, they all can't find the time or money to build an engine from the bottom up, but that doesn't seem to bother them.

As Crawford says: "It's the fine tuning that's the most fun. We've been involved in auto racing for eleven years, but I'd rather go lobster boat racing any day. With a lobster boat, you're in the corner all the time. If you have a weak link, you're going to find it. Last year, I was off by two degrees in my timing and roasted a piston."

How long does it take to roast a piston?

Crawford laughs. "About ten minutes," he says.

The Candle That Burns Twice As Bright Burns Half As Long

Doug Mitchell of Mitchell's Garage in Frankfort, ME knows all about weak links. Last year his 800 hp 460 cubic inch Ford broke a crank at 6,400 rpm. The engine had been taken apart every fall, and still it broke the crank. Fatigue just got the better of it after a grand total of 80 hours, and it was a winning engine, too.

"Anybody can turn an engine hard," says Mitchell. "The trick is making 'em so they stay together. To get horsepower, all the parts have to work in conjunction. It's the combination, the balance of all the little things. That's where you develop you're optimum horsepower."

Glenn Holland of Holland's boat shop in Belfast is quick to agree. This year will be the third time in a row that he's sent the *Red Baron's* engine to Hofstetter. He does this because he thinks that the Florida speed shop can build him the best, most endurable racing engine possible. "Under full race conditions," says Holland, "Hofstetter says his engines will last 100 hours." In fact, one of Hofstetter,s favorites is a 622 cubic inch 950 hp model. "We have a pair of these that are going on 6 years old and 350 hours," he says.

Holland doesn't like to talk too much about the current specifications of the *Red Baron's* engine, which has a Detroit Diesel 12-71 blower on it for a supercharger. But at this stage in the game, he's willing to say that what started as a 460 cubic inch motor is now about 600 cubic inches. He'll also tell you that at 5,900 rpm, it's pushing well over 800 horsepower, a specification that easily places it in the 'burning twice as bright' category.

Meanwhile, people are doing to diesels what they've been doing to gas engines for years and years.

Mitchell says anything you do to get more horsepower out of a gas engine you can do to a diesel. However, as far as he knows, nobody has gone inside them...yet. "They're mostly playing with fuel ratios and turbochargers," he says. "Basically, they're doing the same thing I've done...over-fueling the engine and giving it more air."

According to Holland, the main reason people aren't

fooling around with the inside of their diesels is the expense. A new gasoline 502 might cost a few thousand. A new diesel could cost tens of thousands, and more.

The Dynamometer

The best way to keep a race engine together is to build it of the finest materials, make sure the clearances are at an absolute minimum, then tune it to perfection. But to do the latter, you really need to use something called a dynamometer—a special piece of bench testing equipment that controls the output or driving torque of rotating machinery.

At Chief Engines, dynamometer testing is done on a Superflow 901.

"Friction horsepower," says Hofstetter, "is really neat feedback." In other words, how much power does it take to turn the engine itself? It's always a question to lower that number, because sometimes you can lose as much as 200 hp in internal friction. For example, the 502 loses 121 hp at 5000 rpm."

A perfect example of how dyno testing can vastly improve an engine can be found in the 509 cubic inch model that Hofstetter originally designed and built for a special class of offshore racing boat. He explains it as follows: "The engine had to have a stock cylinder head. You could port it, but not much else."

Some of the alterations that were made to the 509, a $45,000 engine, included: a different type of intake manifold, different air filters, and hand polished valves. Hofstetter says it was a tough job that required running 200 'pulls' or acceleration patterns on the dyno.

But the 509 was a special case. With a model that he already knows, Hofstetter usually spends an average of six to eight hours testing the engine on the dynamometer. If there's going to be a failure, it'll happen in the first couple of hours. "We use the dyno for two reasons," he says, "first, to design and build a more efficient and higher output engine, and second, to make sure the product performs the way we say it will."

Dollars and Sense

Glen Crawford,s pretty proud of the fact that the *Little Jan's* engine started with a total investment of $600. "The guys in the shop joke about it," he says. "They call it my garbage can engine."

Joking is right, because the $600 doesn't even scratch the surface in terms of total funds used to get the motor where it ended-up.

"I spent some money," says Crawford. "I had problems with pistons, so I decided to get away from carburetion and go to fuel injection. The injection system was designed by a company in Texas. That was $4000—the most I ever spent on a lobster boat engine. So now there's no baseline. We have to start all over again, right back to square one. This is a huge jump, but that's the fun of it."

Of course, from the money point of view, Crawford's engine is a bargain. Some of Maine's other racers are spending upwards of $40,000—and that's just for a gas engine. With diesels, the investment starts at 40 grand and goes up from there. For example, rumor has it that one racer will be crossing the starting line in a 38' super-light boat powered by a stock 1,100 hp MAN diesel. The cost of the engine and gear...about $130,000.

SUNSPOT

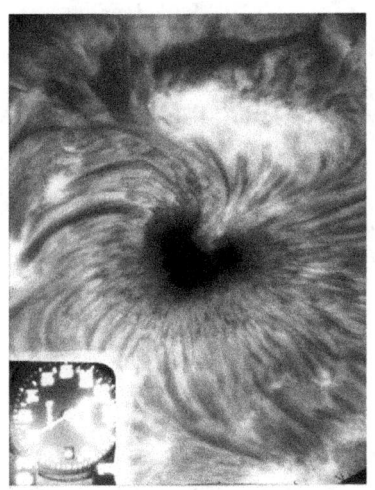

CHAPTER NINETEEN
TRUSTING YOUR ELECTRONICS

GPS Isn't Perfect

Most of today's boats have integrated navigation devices that rely heavily, if not exclusively, on the Global Positioning System (GPS). Radio signals from the network of GPS satellites serve as the electronic foundation for plotters, sounders, sonars, and radars. Without the signal, the GPS receiver wouldn't be able to continually cross-fix a vessel's position, and integrated peripheral equipment would end-up displaying a "data unavailable" error. No signal, no cross fix, no vessel tracking.

Differential GPS and WAAS-enabled DGPS are clearly an improvement over their predecessors, SATNAV and LORAN, in terms of reliability, coverage, and accuracy, but is it infallible?

According to scientists at NOAA's National Geophysical Data Center (NGDC), and its sister organization, the National Space Environmental Center (NSEC), both located in Boulder, CO, the answer is: Not really. Like all other transmission sources, both land-based and space-based, GPS can be knocked-out by extra-terrestrial disturbances.

Minding the Store

The NGDC and the NSEC are two of our government's most active and and least known scientific organizations. They don't have the "sexy" rockets and space shuttles of NASA or the large ships of NOAA's National Ocean Survey, but their job is equally important. Basically, the two centers manage environmental data in the fields of marine geology and geophysics, paleoclimatology, solid earth geophysics, glaciology, and solar-terrestrial physics. In fact, it's the latter that relates to solar activity--and it's solar activity, particularly in the form of solar flares and coronal mass ejections, that can and do affect the functioning of GPS and other marine navigation and communication transmission sources.

Helen Coffey, former head of the Solar-Upper Atmosphere Data Group at the NDGC, points out just how powerful and interactive these solar eruptions are.

"On March 13, 1989," she explains, "there was a magnetic storm caused by a big solar event. When it was over, about six million people were out of power for about nine hours."

Some of these people, of course, were fishermen at

sea. Although they didn't experience onboard equipment malfunctions, they were affected by the interruption of shore-base marine services, e.g. Loran, cell phone, and marine radiotelephone. "We didn't have GPS or DGPS back then," says Quarter Master 2 Bill Schmidt of the 1st District U.S.C.G. ATON (Aids to Navigation) Office in Boston. But, he adds, if we had, they would have been just as susceptible to the blackout as the Loran-C stations.

It's hard to believe a storm originating 100 million miles away can have such disastrous effects, but this is exactly the case. Solar eruptions send out massive amounts of high energy particles and radiation in the form of a "wind." The solar wind, as its called, is always pulsing continually outward from the sun in varying degrees. On Earth, it causes the aurora borealis; it also alters the growth of plants and trees, as evidenced by the study of tree rings. But during high periods of solar activity, when a solar event significantly increases the solar wind's intensity, a lot more is at stake than a pretty light show in the northern sky and some variations in the growth of plants.

"When it [the March 1989 eruption] hit the environment," says Coffey, "it caused the [Earth's] magnetic field to compress. The field got 'pushed in,' and it induced a current in the ground. The current overloaded the circuits. There was a transformer in the U.S. that actually melted. This thing was as big as a house."

Satellite Damage

Solar flares and coronal mass ejections are associated with increases in the numbers of sunspots, and sunspots are cyclic in nature. The cycle--from the end of one period to the start of the next, or from one peak to the next peak--is 11 years. The chart below shows solar activity and predictions. As you can see, year 2000 was at the waning of a peak year, which would place 2011 at about the bottom of the curve, and place us at present time (2014), in a heightening period of solar activity.

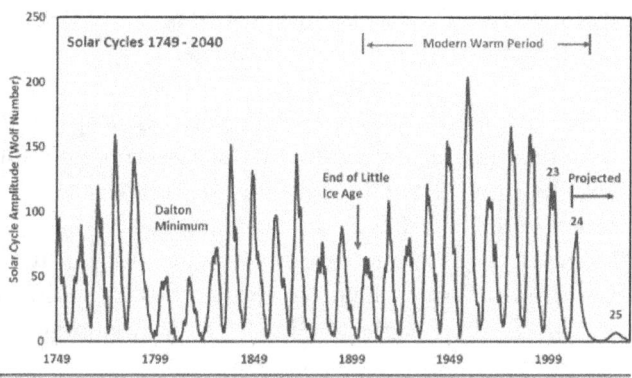

As illustrated by Coffey's example above, a solar event can cause potential problems for a lot of people, which is why scientists at the NGDC and NSEC in Boulder keep a 24 hour watch on the sun, seven days a 2 week, 365 days a year. Their monitoring satellites, A.C.E. and its predecessor, W.I.N.D., give them, and subsequently their customers, about a one hour head start on a major solar event. "We're in communication with a lot of people about solar and magnetic activity," says Joe Kunches, of the NSEC Forecast Office, Space Weather

Prediction Center. "For example, because they're constantly monitoring the integrity of their navigation systems, the U.S.C.G. in Kodiak, Alaska is at the top of our list."

So too is the U.S.C.G. Navigation Center in Alexandria, VA. According to Schmidt, when warnings are issued from Denver, the U.S.C.G. responds as soon as possible by putting out notices to mariners in the form of "pan-pan" broadcasts over the VHF, SSB frequencies, and NAVTEX. "DGPS warnings are issued directly from Virginia," he says. "And for now, we're directed by NavCen to do the Loran-C warnings."

Meanwhile, in the case of a really big eruption, there isn't that much anybody can do. On land, equipment can be shut down or brought to a level of reduced function or operation. In space, it's not that simple. "Circuitry and microchips are at risk," says Ray Conkright, former head of the Ionosphere and Space Weather group at the NGDC. "If the discharge is high enough, they could be affected."

However, although nobody knows for certain-- because GPS satellites didn't exist during the last peak period, and because nobody can say for sure how bad an event we're going to get next year--Conkright and Kunches both believe the GPS birds are sufficiently 'hardened' to withstand the discharge from almost any solar eruption. And yet, while chances are good they won't get fried, they might suffer longevity problems. "Solar cells are always deteriorating," says Conkright. "During a solar event, they might take an extra dose, which could further reduce their life span."

Effects on Radio Transmissions

All of our navigation and communication signals are transmitted as radio waves, and radio waves are dependent in one way or another on the status of the ionosphere, the layer of ionized gas at the very upper end of our atmosphere. Low frequency single sideband and Loran signals bounce off the ionosphere to get to where they're going, and higher frequency GPS, Broadcast, and SATCOM signals shoot through it. This means that changes in the ionosphere can interfere with some of our transmissions.

Think of the ionosphere as an ocean and the discharge of a solar event as a wind, which we've already said it is. The wind hits the ionosphere and creates turbulence. For example: If we're on a real ocean. and it's rough, and we're trying to pick out a radar target from a small boat, the waves would cause interference and false echoes.

In a simplified way, this is what happens when the discharge from a solar event hits the ionosphere. The high energy particles of the solar wind create turbulence that interferes with the echoes and transmissions of radio waves. Fortunately, not all frequencies or all areas are affected equally. During a solar event, higher frequency transmissions like GPS are less susceptible than lower frequency transmissions (Loran, Single Sideband), and the equatorial region is less susceptible than the higher latitudes.

Even more problematic is what happens to the Earth's magnetic field during a solar event. "All of the U.S. Geodesic survey people shut down operations during periods of high magnetic activity," says Conkright.

What this means is that vessels can lose navigation and communication electronics for as long as it takes for the ionosphere and the magnetic field to calm down. How long can can it take to calm down? "A solar event in the northern latitudes can create a disturbance that lasts for two or three days," says Kunches. "HF communications can be blacked out for days. GPS would be hampered for days. Magnetic compasses would be off for days."

It's not a great scenario if you happen to be tracking your vessel's progress or trying to find the shortest route home. For this reason, you want to be schooled in the use of the other navigation tools at your command: the gyrocompass, which uses the spin of the Earth and not the magnetic field to find true north, expensive and not for everybody, the sextant, which is inexpensive and easily available, but more difficult to use, and the chart work kit, and with which includes dividers, parallel rules, and clock, with which every mariner should have complete familiarity.

###

Electronic Charts

Over the years, seafarers have regarded electronic charts with the full range of viewpoints. First there was disdain, then suspicion, then some form of curiosity, and finally, acceptance. As more and more boats began installing the equipment, captains, particularly those on big boats, began to think of electronic plotting as less of a luxury or a convenience and more as a necessity. Today, electronic charts and plotting software are in the "don't leave home without them" category.

These days we get tons of functionality out of our plotters because marine hardware and software manufacturers take full advantage of the electronic chart maker's enhancements and upgrades. Higher resolution, 3-d, better graphics, interconnectivity, and new software and map features for mobile devices, are turning these tools into indispensable electronic aids to navigation. Available for stand alone, proprietary device from a host of marine electronics manufacturers, or for PC, Mac, Tablet and mobile phone users.

Here's a brief overview of what electronic chart makers have done to add features to their products.

C-MAP

According to C-Map in Mashpee, Massachusetts, the company is continually working hard to expand and enhance its product. Over the years, enhancements have included the port database, which basically provides a yellow pages of port facilities from around the world, a feature that allows you to open a window on your screen that will give you a list of chandleries, pipe fitters, etc. Navigation software manufacturers for mariners pioneered this feature, and you now find it everywhere, in your home, auto, and on your phone. And while these databases are only as good as their updates, C-Map is dedicated to keeping information current, with weekly upgrades of all chart systems features.

Nautical Manager is one of the company's newest products, and here's what they have to say about it:

NauticalManager from Jeppesen allows mariners to easily plan, generate, and document routes using an ECDIS and onboard PC. It also makes it simple to create the reports required by port state control and class auditors for checking a vessel's ECDIS compliance.

NauticalManager eases the bridge-side burden of ENC management by making it simple to order Official ENCs, review the chart portfolio and integrate temporary and preliminary Notices to Mariners.

NauticalManager was developed after ECDIS users requested a way to simply or automatically plan routes that are in compliance with the ECDIS mandate. It features automated route planning and chart selection – resulting in better voyage planning and less time spent managing digital information.

Routes are generated from a database containing thousands of waypoints, legs and ports, allowing a mariner to simply edit a recommended route and transfer it straight to an ECDIS. With NauticalManager, users also can create a semi-automatic route from anywhere on a chart.

The NauticalManager software can also integrate supplementary data layers that provide:

1. ENC Notices to Mariners, including daily temporary and preliminary (T&P) notices
2. DNV-recognized weather information
3. Accurate piracy overviews
4. Notification before charts expire
5. Weekly updates of the entire chart portfolio
6. Notification when new charts, weather updates and chart licenses are available

The ability to plan a route according to weather and traffic conditions and with the right charts can increase savings in a competitive shipping market, says Jeppesen's product manager Geir Olsen: "If a ship is detained in port due to a lack of charts or outdated charts, expenses are huge compared to the almost insignificant cost of obtaining the charts."

"In addition to charts, mariners and ship operators need products and services that improve safety and efficiency and save money. Outdated weather faxes directly affect efficiency, for example. Access to data about the latest piracy incidents is also crucial for maintaining security and international ship and port security (ISPS) levels in certain areas."

To increase user friendliness, NauticalManager guides mariners through the process of route planning, chart selection and ordering and the downloading of weather forecasts. The solution also comes with instructions on how to keep a chart database up to date and how to export chart licenses to the ECDIS/ECS. Selecting chart updates is done at the click of a button.

The NauticalManager software is available to Jeppesen customers at no additional cost; users pay only for data, such as weather and piracy information and chart licenses. Piracy information comes through a subscription to Jeppesen Piracy Service, while a three-month trial subscription to the Jeppesen ENC and weather services is included with the software. This provides access to a global ENC portfolio, marine weather forecasts, tropical cyclone warnings and information about oceanic currents and ice conditions.

NauticalManager can be used on most onboard PCs through marine VSAT or other fleet broadband connections and also supports email communication protocol.

To learn more about NauticalManager, Jeppesen's ENC service and innovative solutions for the world shipping industry, visit www.jeppesen.com/main/corporate/marine/commercial/nauticalmanager

NAVIONICS

Navionics claims to have the largest marine and lakes database in the world, and they have taken a similar route in terms of providing more options and faster upgrades for customers. Navionics has also recut their cruising charts in terms of logic, meaning, you can save money and not buy large cruising areas you don;t plan to visit.

As with C-Map, Navionics has continued to upgrade their enhanced features, e.g. full port service coverage, their multi-dimensional data, and their birds eye view library. And they've increased their zoom levels so that screen displays are matched to radar levels. For example, this means you can pick a zoom level that matches the 1.5, 3, 6, 12 or whatever mile range you've chosen to view.

For a list of Navionics combatible devices go see the Navionics Compatibility Guide.

MAPTECH

Maptech, Inc. continues to upgrade and enhance its Photo Regions and aerial photographs of port entrances and approaches. "Photo regions have been out since the the late 1990s and started out with Florida, then the Northeast, Delaware Bay and Jersey. Chesapeake Bay, Puget Sound, Southern California and the San Francisco Bay soon followed. It's now virtually complete.

For those who've never seen the photo region feature, an example is as follows: Coming into Government Cut in Miami, there's an aerial shot of the channel itself, and just beyond, a different angle showing the Miami Harbor marina complex. In this case, the plotter would show various approaches in full color, each one date stamped. The date stamping is important because some areas might have dramatic seasonal changes that could be confusing, i.e. a snow-covered hill instead of one covered in summer green.

Unlike NAVIONICS, C-Map, and SOFTCHART, Maptech uses regular NOAA charts, which are regularly updated in cooperation with NOAA, NEMA, the Canadian Hydrographic Service, and the US Coast Guard.Updates on a subscription bases are available to mariners who have purchased the professional version of the Maptech software. The company also has a half price "Lite" version of their plotting software which may be sufficient for recreational boaters. All products can be downloaded to PCs from the company website: www.maptechnavigation.com

Chart Navigator ($249) comes with complete U.S. coverage of Raster charts, satellite photos, aerial photos, and topos. Also included is the full-featured Offshore Navigator software program, which provides the tools for route and trip planning and navigating in real-time, plus tides and current information for U.S. waters. Chart Navigator can interface with autopilots and other NMEA0183 instruments.

Maptech's new U.S. Boating Charts with Tides and Currents on DVD ($99 for PC and $199 for Mac) has over 2200 up-to-date NOAA nautical charts and more than 750 river charts from the Army Corp of Engineers. It comes with GPS real-time navigation software.

Chart Navigator Pro ($499) features: ALL Raster NOAA Nautical Charts, Contour 3D, Photo, Topos and more. Coverage - ALL U.S. coastal areas and major river systems on 18 DVDs. No charge telephone tech support and Chart Updates starting at only $49.95!

Maptech has designed new color pallets for daylight or night viewing, and the company at one point was working on vector charts that conform to the S-57 Standards. S-57 (Edition 3) refers to the new standard for vector charts, also known as Electronic Navigation Charts (ENCs), set down by both the International Hydrographic Organization and the International Maritime Organization and scheduled to be adopted in 2010. In the near future, possibly this year, 2014, the only way a captain or vessel owner will be able to have a paper-less wheelhouse is if his or her equipment complies with the S-57 standard.

NOAA's goal was to first produce the 40 deep water harbors in the vector format. Eventually, they filled in the rest of the US coast. As charts continue to roll out, you may start to see a hybrid of two products, raster and vector. Here's the latest from NOAA-ENC:

NOAA_ENC

NOAA Electronic Navigational Charts (NOAA ENC®) are vector data sets that represent NOAA's newest and most powerful electronic charting product. NOAA ENCs conform with the International Hydrographic Office (IHO) S-57 international exchange format, comply with the IHO ENC Product Specification, and are provided with incremental updates that supply Notice to Mariners corrections and other critical changes. NOAA ENCs and updates are available for free download. NOAA ENC data may be used to fuel Electronic Chart and Display Information Systems (ECDIS).

NOAA ENC is a vector database of chart features

built to the IHO's S-57 standard. NOAA's Office of Coast Survey, as the U.S. national hydrographic office, is exclusively responsible for production and authorization of NOAA ENC data in U.S. waters.

NOAA ENCs support all types of marine navigation by providing the official Electronic Navigational Chart used in ECDIS and in electronic charting systems. NOAA ENCs support real-time navigation, as well as the collision and grounding avoidance needs of the mariner, and accommodate a real-time tide and current display capability that is essential for large vessel navigation. NOAA ENCs also provide fully integrated vector base maps for use in geographic information systems (GIS) that are used for coastal management or other purposes.

Building and Maintaining NOAA ENCs

In 1997, NOAA began a process of building a portfolio of ENCs that encompass the same areas covered by NOAA's suite of approximately 1,000 paper and raster charts. The ideal and most accurate way to build ENCs is to recompile the paper chart from all of the original source material. Unfortunately, this process is impractical as it is far too labor intensive. Instead, NOAA ENCs have been compiled from source on those features that are deemed to be navigationally significant. U.S. Army Corps of Engineers' federal project limits have been captured from large-scale drawings. These precise coordinates of channel limits are being incorporated into the ENC. Likewise, high-accuracy positions are being used to chart U.S. Coast Guard aids to navigation. The paper chart has been the source for the remainder of items.

NOAA has utilized private contractors to build NOAA ENCs. Private companies are provided high-resolution source information such as U.S. Army Corps of Engineers channel limits, and aids to navigation established by the U.S. Coast Guard. Contractors are also provided with the latest version of the paper/raster chart. All NOAA ENCs that are built by private contractors are reviewed by NOAA cartographers before they are posted on the Internet.

NOAA cartographers and private contractors (under NOAA supervision) apply updates to ENCs using high resolution original source material. As new source information arrives at NOAA headquarters, cartographers update NOAA ENCs using high resolution position and depth information.

Building Status

As of January 2013, 965 NOAA ENCs are available for download. All of the major ports throughout the country now have NOAA ENC coverage. Many smaller scale coastal ENCs that connect these ports have been completed, while some are still in the building process. NOAA plans to continue its ENC building program in the upcoming years.

Updates for NOAA ENCs

NOAA's goal is to provide weekly updates for each ENC it makes available. Update cells are posted on the

Internet for download. When downloaded, these update cells can be applied to the base ENC cell to produce an up to date ENC.

Distribution of NOAA ENCs

NOAA ENCs are available as free downloads from the Internet. Mariners who wish to download NOAA ENCs directly and use the data to fuel ECDIS or ECS may do so.

In October 2005, NOAA created a new mechanism for distributing NOAA ENCs. NOAA announced certification requirements with standards for applicants who want to redistribute NOAA ENCs as official data. Two types of certification are offered. The first type, Certified NOAA ENC Distributor (CED), covers NOAA ENC downloading, exact copying, and redistribution of those copies. The second type, Certified NOAA ENC Value Added Distributor (CEVAD), permits reformatting official NOAA ENCs into a System Electronic Navigational Chart (SENC) using type-approved software, and distribution of that SENC. NOAA intends by this action to assure that, though redistributed, quality official NOAA ENC data is offered to the public in support of safe navigation on U.S. waters.

In order for an electronic chart to gain type approval as an ECDIS, it must be fueled by official ENC data. However, ENC data is not only for ECDIS use. ENCs can fuel any ECS that reads the S-57 format. Private vendors are free to download NOAA ENC data, reformat it into a proprietary format, and then resell that data.

Free ENC viewers are available online. Links can be found at the NOAA-ENC web site.

Softcharts

Softcharts was bought out by Maptech, and the company has recently announced it is discontinuing the Softcharts product.

###

Plotters

In the modern world of navigation, software is king. Without computer code, your electronics are nothing more than tiny bits of melted and molded sand and metal. In fact, even your raster scan radar, GPS, and video sounder need a preprogrammed silicone chip to make them work.

Remember the old days, thirty-some years ago? Remember all the guys who would scoff at the newfangled electronics? They'd point to those big Loran-A units and say things like: "If you need that whatchamagig, you shouldn't be out there." Well, not anymore. Today, if a mariner isn't on the technology wagon, he or she is left behind, or worse.

Meanwhile, for marine-oriented software manufacturers, a lot of the action is taking place in the area of mapping and plotting. At least half a dozen companies are advertising packages for mariners, and several are writing code just for commercial fisherman.

Here's an overview, and some choices the shopper faces.

Designated Plotters Versus PC-based Plotters

What's better? A software package designed to run on a personal computer, e.g. Macintosh or Windows environment, or a designated plotter, Furuno, Northstar, Raytheon, Sitex, etc., something that comes preprogrammed with its own dedicated buttons? The answer is simple: It depends on your application, economics and boat.

Designated plotters don't have the horsepower or capacity of a full-blown PC Plotter. They may not have the speed, versatility or memory. They can't be used for other things like writing logs, doing accounting, or connecting to the internet. And, except for the very high-end and expensive designated plotters, their display screens can be rather small. On the other hand, what they lack in power. they more than make up for in terms of ease of use and dependability. Whereas, in most cases, you can take a designated plotter out of the box, plug it in, and start using it, a PC-based plotter involves a lot more work setting-up and operating.

For one thing, an off-the-shelf PC can be kind of fickle about where and how it lives. It needs a 'clean and stable power source, as well as a well-ventilated, relatively cool and dry operating environment. It needs to be a safe distance from any magnetic influences, such as inverters or motors with copper windings. And, according to software manufacturers, it should be isolated from vibration. 'A standard desktop [PC] won't hack it unless it's soft mounted.

Ideally, your desktop PC should be soft mounted on rubber or foam in a dry wheelhouse. It also should be wired securely. Steady power is a real critical issue, and most pros will recommend a 660 Volt-Amp UPS. It has sufficient power, and it will give you thirty minutes to shut things down if something goes wrong. This means you don't have to worry about the computer if you're switching from one genset to another.

Does this make the laptop, with its built-in battery, a good choice? The answer is a resounding yes. The laptop's internal battery kind of categorizes it as a computer with its own UPS, i.e. the battery automatically takes care of spikes, and they have smaller keyboards, track pads, and screens. If you want, you can add larger peripherals, but doing so just defeats the logic for getting a laptop in the first place.

Meanwhile, several after-market companies build Desktop and Laptop PCs specifically for industrial and marine applications? These computers have water-resistant, and, in some cases, waterproof keyboards. They also have moisture resistant circuitry and super-dampened drives. But even though they're much tougher and more resilient than a standard, off-the-shelf PC, they still have two Achilles heals: they need to be cooled by drawing air over their circuit boards, and they have a hard drive for storing data. Consequently, vibration and temperature extremes should still be avoided.

By comparison, the installation of a designated plotter

is a plug and play procedure. Simply provide 12 Volt power and NMEA input from your GPS, radar, sounder, etc., and you're set to go. The units themselves are already hardened for marine use, because they're built by marine electronics companies and not office machine companies, and they're getting more and more powerful all the time. Designated plotters can be integrated with fishfinders, radarS, autopilots, etc., and new features are being added every year .

Horsepower Does Not Mean Get Up and Go

It's a catch-22. The marine software industry wants to give the customer the most bang for it's buck, but at the same time, they want a machine that's easy to use. There's no question that a full-function PC requires a greater amount of learning time than a designated plotter. Not only do you have to learn the plotting software, but you have to be somewhat familiar with the computer's operating system. Whether it's Windows. MAC/OS, IOS or Android or something else, a user has to invest in some learning time. Unfortunately, sitting at a desk or a computer is not what a lot of mariners want to do.

As one retailer told me: "I've had guys call me from the Bering Sea and ask: "Hey, how do you turn this thing on?"

In other words, different customers have different levels of computer experience. By the same token, they also want different things from their marine electronics. Most professional mariners will only learn what they have to know in order to make their device do the things they want it to do. Learning to use all the functions and features of a given piece of equipment sometimes has a diminishing law of return.

This is typical for computer users in general, and, for all intents and purposes, it's the proper way to make the most out of a software package. Learn what you need to know to do the job, then learn more as both the demands of the job and your familiarity with the software increase.

Of course, the same applies to designated plotters. They have easier-to-use keypads and simpler scroll-down menus, but you still can't start one up without reading the manual and doing some homework.

Features for the Future

When it comes to features for mariners and particularly fishermen, not all plotters are created equal. For example, in Alaska, a lot of captains like to be able to import ARPA data. Plotting ARPA (Automatic radar Plotting Aids) data lets a captain know what the other boats around him are doing. In other words, not only can he track the progress of his own vessel, he can track the progress of 30 or 40 additional targets. Obviously the immediate benefit of this is one of collision avoidance. However, for fishermen, the real advantage is being able to make a tow where, as Captain James T. Kirk might say: "No one has gone before."

Clearly, if a fishing trawler captain wants an edge, he

should be certain the plotter he buys will track and plot radar targets. He should also be sure his plotter will work with bathymetric charts. The bathy chart shows seabed contours, and I don't have to say how important these are to a fisherman.

By the same token, Gulf of Mexico fishermen might want to consider oil block charts. Shrimpers, for example, buy a lot of oil block charts.

What's next for plotting software? More integration. More 3-d rendering. C-Map has something it actually calls 4-d. I'm not sure exactly what that is, but if it takes you to another dimension, I'm in.

CHAPTER TWENTY
Navionics Compatibility Guide

Untitled

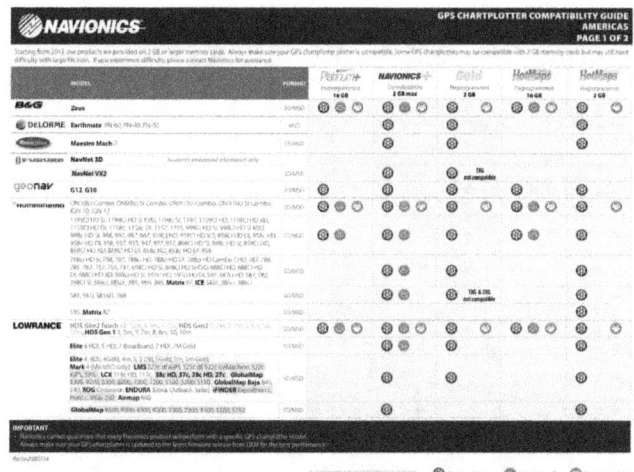

###

Untitled

Robert Bernstein

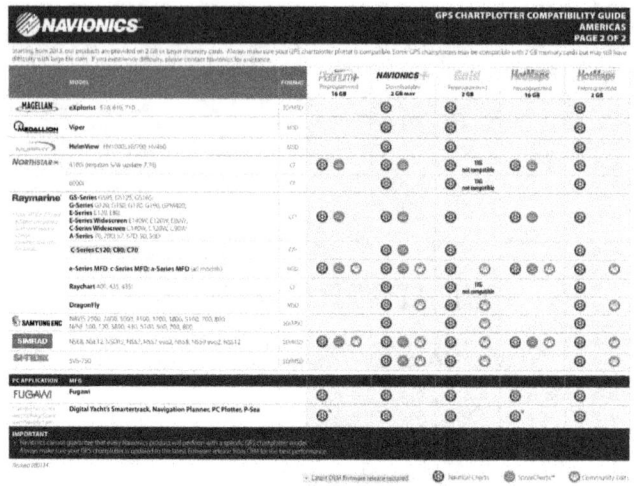

WORLD'S SMALLEST 70 LPH WATERMAKER BY HP

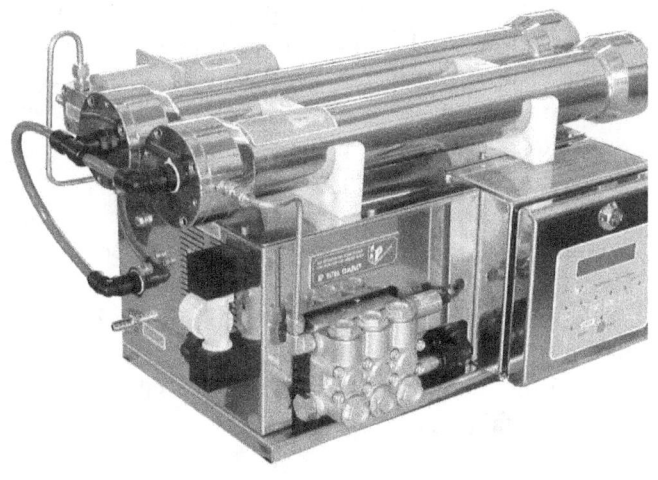

CHAPTER TWENTY-ONE
WATERMAKERS

Water, Water Everywhere, Nor Any Drop To Drink.... Hah! The crew of the Russian factory ship, Suverinitet, would spit in the face of that sentiment. Because when Samuel Taylor Coleridge penned The Ancient Mariner in 1798, the idea of a boat carrying a 350 ton per day watermaker was as farfetched as a trip to the moon.

The Suverinitet—one of 15 Russian vessels built in Spain for head-and-gut-type fish processing—was recently refitted as a Surimi factory ship. Most of the work was done at the Victoria Shipyard in Victoria, British Columbia, but the detailed job of designing and supervising the installation of the desalination system went to Alfa Tec of Seattle.

"It takes roughly 2-1/2 tons to 5 tons of water to produce 1 ton of Surimi, depending on the ship's processing room," said John Case, Alfa Tec's General Manager. "For the Suverinitet, we chose an Alpha Laval rising film evaporator."

Case admits he's somewhat partial to evaporators. Speaking from Dutch Harbor, Alaska, where he was finishing up with the Suverinitet and checking on his other customers, he pitched a pretty good argument for choosing evaporators over reverse osmosis watermakers. "They might be more expensive to purchase," he says, "but they produce better water, and they're cheaper and more reliable to run." However, he's quick to point out that the smallest vessels he installs them on are in the 75' range. The reason for this is two-fold: Evaporators take up more space, and they require a waste heat source.

Frank O'Hara of the O'Hara Corporation in Rockland, Maine, agrees. His company switched all three of their boats from reverse osmosis to evaporator-type watermakers. Why? "Cost," he said, "In one year, we spent enough on membranes to justify the upgrade."

Still, O'Hara admits that when the company first converted the boats and sent them to Alaska, it couldn't afford the more expensive evaporators. In other words, reverse osmosis was the only alternative at the time. It got the boats out fishing and making money. And it is the the type of watermaker chosen by yachtsmen nine times out of ten. The main reasons for this are its size and design. Making water with an evaporator is a lot more complicated.

Reverse Osmosis (RO)

Watermakers don't actually make water. That would

take something akin to the replicators aboard Gene Roddenberry's Starship Enterprise. What watermakers do is generate or convert water by removing salt, or, more accurately, dissolved and suspended solids. Given that seawater averages 35,000 part per million of total dissolved solids (ppmtds), and taking into consideration the World Health Organization's drinking water standard of 500 ppmtds, it's clear these devices—watermakers, desalinators, water generators, water converters, or whatever you call them—have their work cut out for them. Other methods of making water—freezing, electrodialysis (via an ion exchange membrane), and piezoelectrics—are just not practical for use aboard boats.

Reverse Osmoses (RO) is based on a process that didn't have a name until Scottish chemist Thomas Green coined one in the middle of the nineteenth century. But regardless of his timing, the process of osmosis dates back to the beginning of life. In fact, solids have been filtering in and out of membranes ever since the first single cell organism took a proverbial backstroke across the primordial pond.

In an RO watermaker a high pressure (800 to 900 psi) pump forces intake water through a tubular, thin-film composite membrane with holes 5 to 10 million times smaller than a grain of sand. The membrane acts as a filter, only allowing certain atoms and molecules through while "filtering" out solids, bacteria, viruses, and heavy metals. The force required to stop the flow from the dilute side to the solute side of the membrane is called the "osmotic pressure." Picture two vertical tubes separated by the membrane, seawater on one side, the dolute side, fresh on the other, the solute side. If we increase the pressure on the seawater side to more than its osmotic pressure (355 psi for seawater), the process of osmosis is reversed and H20 molecules cross the membrane to the fresh water side; in fact, ROs function at more than double seawater's osmotic pressure in order to attain peak efficiency.

As mentioned, in terms of an RO watermaker's size and power requirements, there are a lot of advantages over other types of watermakers. But there are also disadvantages. For one, the osmotic pressure rises as the concentration of filter material increases, and at some point the concentrate has to be removed, which means changing the membrane. This is a big part of the RO watermaker's scheduled maintenance, and the cost is not inconsiderable. For example, a 2-1/2" X 40" membrane for a 500 gallon per day (gpd) watermaker can run anywhere from $250 to around $650, depending on the brand and model. On a yacht, it's something that might have to be replaced every six months. On a commercial vessel, even more often. And a lot of this maintenance depends on where you're making water, out at sea, where there are fewer contaminants, or closer to shore, where there are more.

Evaporators

Evaporators have been around since c.49 B.C. when Roman legions used large open pools to distill fresh water from the Mediterranean. Nowadays, the equipment is a lot more sophisticated, to say the least. A modern evaporator gets its heat, not from the sun, but from electricity, a boiler, engine jacket water, exhaust gas, or a combination thereof. Systems are balanced and matched for optimum efficiency, and they come with their own confusing terminology. To put it simply: If this were Evaporator Jeopardy, the general category would be single-or multi-effect distillation (MED), and the four subcategories would be——Flash, Thin-Film, Submerged Tube, Plate, and Vapor Compression.

Of the over 7,500 shore based systems in the world, roughly 2,000 are flash-type, 3,000 are other MEDs, and 1,500 are RO plants. One MSF (Multi-Stage Flash) facility in Saudi Arabia produces 250 million gallons per day. In marine applications, MSF systems are on yachts, fish boats, cruise liners and Navy ships. "We have 100,000 gallon per day MSF evaporators on several carriers," said Philip Liu, a mechanical engineer with Beaird Industries in Louisiana.

Multi-stage Flash evaporators operate by vaporizing seawater in a specialized chamber. The brine——its vapor point lowered by a vacuum——flashes repeatedly. In other words, hot seawater is injected into a chamber where the pressure is lower than the vapor pressure of the water. This pressure difference causes the water to flash, which means it goes from liquid to steam without boiling. The vapor is then collected and passed through a condenser, at which time the distillate is collected .

In evaporator terminology, the word "stage" refers to the number of times the process is repeated. Add a stage or effect (the words are interchangeable) and you get a corresponding increase in efficiency. In other words it takes less energy to make more water. However, when you add a stage you also add equipment, weight, etc. A shore based plant might have 20 stages or effects, a ship 1 to 4. For example, Alpha Laval's rising film evaporator ——one of the more efficient according to John Case ——needs 32 kW to make 1 ton of water in a single stage configuration, 16 kW in a 2 stage configuration, and 8 kW in a 4 stage configuration.

The other MED evaporators work in a similar manner, but instead of "flashing" the water, they boil it, and they utilize a more continuous stream of feed water.

As its name implies, the Vapor Compression system—the most efficient evaporator of all—does one thing the others don't. It utilizes an electrically powered rotating compressor to boil the feed water, which makes it a very efficient but expensive unit.

Finally, ST and Plate systems are thermodynamic cousins. They operate on a similar principle by using heat exchangers in direct contact with the feed water. Both provide very good water, but of the two, the plate-type is more efficient. However, because the ST system is the smallest on the market, it can be found aboard vessels with power to spare and hardly any free space e.g., Trident submarines.

Why Install One?

The math is incredibly simple. Let's say you're building a 65' sailboat, or a sportfisherman of similar size. You know the owner's plans will include trips from New York to Bermuda to Grand Cayman, with a total of eight passengers. While it's true that most people around the world are living on less than a gallon of water per day (gpd), the average person in this country uses 3 to 7 gpd, and some overindulgent guests will use as much a 20 gpd. This means that on a trip of ten days, 8 people will need 500 to 1000 gallons. Given that one gallon of water weighs 8.3 pounds and takes up 134/1000 cubic feet, 500 gallons of water will weigh 4,150 pounds (not counting the tank) and take up 67 cubic feet of space. That's a tank of water a little over 4' on a side, weighing more than a full size pickup truck.

Why carry all that water if all you really need is 100 gpd? Why not just install a 150 gallon tank and buy a 150 gpd watermaker? An RO system with this capacity will be about the size of a microwave oven and weigh 50 to 100 pounds. Needless to say, the weight savings is substantial. In fact, a 150 gpd RO watermaker will probably be small enough to fit under a berth in the cabin.

Choosing and Installing One

RO watermakers base their output capacities on a feed water temperature of 77 deg. F. As the intake water temperature drops, so does the efficiency of the watermaker. This means, for example, that an Alaskan crab boat manned by a crew of 5 or 6, working in 39 deg. seawater, will need an RO watermaker with a higher rating. If the requirement is for 500 gpd, they should be thinking about an 800 gpd watermaker; Some companies sell preheaters for this sort of thing, but a lot of professionals won't recommend them.

Another point to consider relates to the Parts Per Million Total Dissolved Solids (ppmtds). As sated above, the WHO recommends drinking water with no more than 500 ppmtds. Seawater averages 35,000 ppmtds. An evaporator produces a distillate while an RO watermaker produces a solute. Translation: There are more ppmtds in water produced by an RO watermaker, and its possible that if the water being produced is used for a boiler, it could cause scaling. For example, an MSF evaporator produces water with less than 6 ppmtds, and the distillate from some other MED evaporators has less than 2 ppmtds. And RO watermaker operating with a new membrane and at peak efficiency will producer water with about 20 ppmtds. In actual usage, however, this problem seems to be more theoretical. ROs watermakers have been supplying water to boilers without problems for many years.

Meanwhile, Scott McGuire, President of Filtration Concepts in California, doesn't see it as an either-or situation. "I won't try to talk somebody out of an evaporator if they've got the waste heat, but there are applications where both make sense. Some boats have the waste heat when they're underway, but when they shut down, or idle, the heat's gone.

As Case explains: "A 400 to 4000 gpd evaporator can be run exclusively off engine jacket water. Over 4000 gpd and the jacket water's boosted with steam from a boiler. You can have an installation where you're drawing heat from two mains and three generators. You can use engine jacket water, exhaust boilers, a steam boiler, whatever, dump it all into a common loop and control the temperature with one or two, large 3-way thermostatic valve(s)." Either way, you need the heat, and without generators and/or boilers running, your water will have to come from an RO watermaker.

Lastly, not all ROs are created equal. Some cannot be rebuilt, and some require more hands-on maintenance. Case and others suggest that buyers research a watermaker company and seek out testimonials from other boat and fleet owners. A good rule of thumb is to look for a manufacturer or reseller that has installations in your specific type of vessel. Bottom line: Choosing and installing an RO watermaker is a lot more complicated than getting a new diesel.

Maintenance Schedules

"We have [reverse osmosis] systems that have been running for seven years without a membrane change," says McGuire, but adds, "We had one guy who was in the Caribbean and wanted to use his RO desalinator to filter the marina's water. He used a black hose and left it lying uncoiled on the dock, under a hot sun. The temperature in the hose reached over 135 deg. F. When the customer ran the watermaker, the hot water burned out the membrane."

Membranes are pretty finicky. They don't like chlorine or feed water over 120 deg. F. They don't like like sand, dirt, seaweed, kelp, rust, oil, etc. RO watermakers should have only non-ferris metals on the intake side, and, in some cases, multiple pre-filters. Most importantly, the membranes shouldn't be dried out. Whenever the unit's shut down, it should be flushed and pickled.

Also, according to manufacturers, the system should be treated chemically once a year, or whenever output or water quality drops below a certain level, typically 15%. "You use a low PH solution for scale and a high PH solution for bacteria." said McGuire. "But if you're making great water, it's really not necessary."

Evaporators also need chemical treatments. MED units get cleaned with acid and a neutralizer every six months or so. And, periodically, they should be brushed clean of accumulated scale. Some MSF units are bathed continually in a one PPM acid solution. These MSF units get a treatment with a stronger chemical once a year.

As far as installations are concerned, the plumbing is pretty straightforward for both types of watermakers: A through hull for the feed water, one for waste water, and a line leading to the fresh water tank(s). RO watermakers are more complicated in terms of setting up pre-filters, an unnecessary addition for an evaporator because the solids are left behind in the distillation process. And evaporators are more complex in terms of utilizing waste heat. With regard to UV or Ozone sterilization...only on the output side of the water tank, the side that leads to the ship's domestic service pumps.